Maurice Norton Miller

Practical Microscopy

A Course of Normal Histology for Students and Practitioners of Medicine. Second

Edition

Maurice Norton Miller

Practical Microscopy
A Course of Normal Histology for Students and Practitioners of Medicine. Second Edition

ISBN/EAN: 9783744689618

Printed in Europe, USA, Canada, Australia, Japan

Cover: Foto ©Andreas Hilbeck / pixelio.de

More available books at **www.hansebooks.com**

PRACTICAL MICROSCOPY

A COURSE OF NORMAL HISTOLOGY

FOR

Students and Practitioners of Medicine

BY

MAURICE N. MILLER, M.D.,

Director of the Department of Normal Histology in the Loomis Laboratory, University of the City of New York

ILLUSTRATED WITH ONE HUNDRED AND TWENTY-SIX PHOTOGRAPHICAL
REPRODUCTIONS OF THE AUTHOR'S PEN DRAWINGS

Second Edition

NEW YORK
WILLIAM WOOD & COMPANY
1891

COPYRIGHT BY
WILLIAM WOOD & COMPANY
1887

The Carton Press
171, 173 Macdougal Street, New York

To

ALFRED L. LOOMIS, M.D., LL.D.,

PROFESSOR OF PATHOLOGY AND PRACTICE OF MEDICINE
MEDICAL DEPARTMENT UNIVERSITY OF THE
CITY OF NEW YORK, ETC., ETC.,

This Volume is Inscribed

BY THE AUTHOR.

PREFACE.

THIS volume has been prepared with a view of aiding the instructors and students of the laboratory classes which are under my direction.

It is also presented with the hope that it may be useful to other instructors.

Again, students often wish to continue microscopical work during the interim of college attendance; to such, it is my belief, these pages will have some value.

Still again, very many practitioners, not having had, during pupilage, advantages equal to those provided by the modern laboratory equipment, wish to acquire more knowledge of microscopy, for its value in practical medicine. To such workers, also, I desire to be useful.

So much technique has been introduced as has been found to be of absolute necessity, and no more. The processes for the preparation and exhibition of tissues are generally simple and always practicable.

In the description of organs, I assume the student has a fair knowledge of gross anatomy, but knows nothing of histology. The scheme or plan of the structure is first described —using diagrams where requisite to clearness—after which the mode of preparing the sections indicated, and, under practical demonstration, every histological detail tabulated in proper order. The drawings will, I believe, aid in the recognition of such elements in the field of the microscope.

The illustrations are exact reproductions, by photography, of my own pen-pictures; and distinction must always be made between the drawings which are schematic—used to emphasize the plan of structures—and those drawn from the tissue as seen in the microscope.

Our literature abounds in excellent works for the advanced student, and this volume is designed to pave the way for their appreciation.

I desire to record my high appreciation of the aid of Drs. Charles T. Jewett, Egbert Le Fevre, E. Eliot Harris, Milton Turnure, H. Pereira Mendes, J. Gorman, A. M. Lesser, J. Alexander Moore, Robert Roberts, Esq., Warden, and Mr. John Burns, Clerk of Charity Hospital, in facilitating my access to valuable tissue for the illustrations and for my own studies.

My thanks are due my First Assistant, Dr. F. T. Reyling, for his indefatigable efforts in furthering the work; and to Mr. A. J. Drummond, for photographical favors.

MAURICE N. MILLER.

NEW YORK, June 1st, 1887.

CONTENTS.

	PAGE
Title-page, Dedication, List of Illustrations, and Table of Contents,	i. to xv.

PART FIRST.

TECHNOLOGY.

THE LABORATORY MICROSCOPE.

Description of the Stand,	1
Lenses,	3
General Adjustment,	4
Adjustment for Illumination,	4
Adjustment for Focus,	5
Method in Observation,	6
Conservation of the Eyesight,	7
Magnifying Power,	7
Measurement of Objects,	8
Sketching from the Instrument,	9

PREPARATION OF TISSUES FOR MICROSCOPICAL PURPOSES.

Tissue Teasing,	9
Section Cutting,	9
Free-hand Section Cutting,	10
Cutting with the Stirling Microtome,	12
Cutting with Schrauer's Microtome,	12
Cutting with the Author's Microtome,	14
Sharpening Knives—Honing and Stropping,	16
Paraffin Soldering,	18
Tissue Hardening,	20
Rapid Hardening with Alcohol,	21
Hardening with Müller's Fluid,	22
Hardening with Chromic Acid,	22
Decalcifying and Dissociating,	23
Imbedding and Infiltrating with Bayberry Wax,	23
Imbedding and Infiltrating with Celloidin,	24

	PAGE
Freezing,	24
Staining Methods in General,	25
Staining with Hæmatoxylin,	25
Staining with Hæma. and Eosin,	25
Staining with Carmine,	26
Staining with Carmine and Picric Acid,	27
Staining with Silver Nitrate,	28
Cleaning Slides and Cover-glasses,	31
Mounting Methods,	32
Labelling Slides,	33
Care of the Microscope,	34

PART SECOND.

STRUCTURAL ELEMENTS.

PRELIMINARY STUDY.

Form of Objects,	35
Movements of Objects,	36
Extraneous Substances,	37

CELLS.

Cell Distribution,	40
Variation in Cell Forms,	40
Flat Cells,	41
Squamous, Stratified, and Transitional Epithelium,	41
Pavement Epithelium,	42
Columnar Epithelium,	44
Ciliated Columnar Epithelium,	45
Spherical Cells,	47
Red Blood-Corpuscles,	47
Blood Plates,	48
Colorless Blood-Corpuscles,	49
Polyhedral Cells,	49
Stellate Cells,	50
Polar Cells,	50

CONNECTIVE (FIBROUS) TISSUES.

White Fibrous Tissue,	51
Yellow Elastic Tissue,	52
Adipose Tissue,	54

CARTILAGE.

Hyaline Cartilage,	55
Elastic Cartilage,	56
Fibro-elastic Cartilage,	57

	PAGE
BONE,	58

SPECIAL CONNECTIVE TISSUES.

Adenoid Tissue,	61
Neuroglia,	61
Embryonic Tissue,	62

MUSCULAR TISSUE.

Non-striated Muscle,	62
Striated Muscle,	63
Cardiac Muscle,	65

BLOOD-VESSELS.

Arteries,	66
Capillaries,	67
Veins,	67

PART THIRD.

ORGANS.

THE SKIN.

Layers or Strata,	68
Hairs,	71
Sudoriferous Glands,	72
Sebaceous Glands,	74

THE TEETH.

The Pulp,	78
Dentine,	78
Enamel,	79
Crusta Petrosa,	80
Practical Demonstration,	81

THE STOMACH AND INTESTINES.

General Histology,	83
The Stomach,	84
Practical Demonstration,	87
Small Intestine,	88
Practical Demonstration,	92

THE LUNG.

Bronchial Tubes,	94
Coats,	95
Practical Demonstration,	97
Pulmonary Blood-vessels,	99

	PAGE
The Pleura,	100
Pulmonary Alveoli,	101
Practical Demonstration, Lung of Pig,	103
Practical Demonstration, Human Lung,	105

THE LIVER.

General Scheme,	106
The Portal Canals,	108
The Lobular Parenchyma,	109
Practical Demonstration, Liver of Pig,	110
Practical Demonstration, Human Liver,	113
Practical Demonstration, The Portal Canals,	114
Practical Demonstration, The Lobular Parenchyma,	116
Practical Demonstration, Origin of Bile Ducts,	118

THE KIDNEY.

General Description,	120
The Tubuli Uriniferi,	122
Blood-vessels,	124
Practical Demonstration, with Low Power,	128
Practical Demonstration, The Cortical Portion,	130
Practical Demonstration, Medullary Portion,	133

THE GENITO-URINARY TRACT. URETER, BLADDER, UTERUS, VAGINA, ETC.

General Histology,	135
Practical Demonstration, Uterus and Vagina,	136
Practical Demonstration, Pelvis of the Kidney,	140
Practical Demonstration, The Urinary Bladder,	141
Practical Demonstration, Urinary Deposits,	143

THE OVARY.

Practical Demonstration, Adult Human Ovary,	144

DEVELOPMENT OF THE OVUM.

Practical Demonstration, Fœtal Ovary,	147

THE SUPRA-RENAL CAPSULE.

Practical Demonstration,	151

SALIVARY GLANDS. PANCREAS. PLAN OF GLAND STRUCTURE.

Typical Glandular Histology,	153
Tubular Glands,	153
Coiled Tubular Glands,	154
Branched Tubular Glands,	154
Acinous Glands,	155
The Parotid Gland,	157

	PAGE
Submaxillary Gland,	158
Practical Demonstration, Parotid Gland, Submaxillary Gland, Pancreas,	157

THE LYMPHATIC SYSTEM.

General Description,	161
Lymph Channels,	162
Practical Demonstration, Lymph Channels of Central Tendon of the Diaphragm,	163
Lymphatic Nodes or Glands,	167
Practical Demonstration, Mesenteric Lymph Node,	169

THE SPLEEN.

Scheme of Organ,	173
Practical Demonstration,	174

THE THYMUS BODY.

General Description,	177
Practical Demonstration,	178

THE NERVOUS SYSTEM.

Structural Elements,	180
Nerve Fibres,	180
Nerve Cells,	181
Connective Tissue of Nerve Trunks,	182
Neuroglia,	183

THE SPINAL CORD.

General Description,	186
Practical Demonstration, Cervical Spinal Cord,	187

THE BRAIN.

Membranes,	191
Practical Demonstration, Cerebrum,	192
Practical Demonstration, Cerebellum,	195

MISCELLANEOUS FORMULÆ, ETC.

Dammar Mounting Varnish,	197
Xylol Balsam,	197
Varnishes and Cements for Ringing Mounts; Dammar Varnish; Zinc Cement; Aniline Colors; Oil Colors; Black Varnish; Shellac Varnish,	197
Preservative Fluid,	199
Normal Salt Solution,	199
Razor-Strop Paste,	199

Silver Staining Solution,	200
Osmic Acid Solution,	200
Weigert's Staining for Medullated Nerve Tissue,	201
Bayberry Infiltrating Method,	201
Karyokinesis,	202
Fixing and Staining Corpuscular Elements of Blood,	203

LIST OF ILLUSTRATIONS.

FIG.		PAGE
1.	Laboratory Microscope,	2
2.	Course of Light Through the Microscope,	3
3.	Free-hand Section Cutting,	10
4.	Stirling Microtome,	12
5.	Method of Imbedding with Pith, Turnip, etc.,	13
6.	Section Cutting with Stirling Microtome,	14
7.	The Schrauer Microtome,	14
8.	The Author's Laboratory Microtome,	15
9.	Method of Honing Razor,	17
10.	Turning the Razor on the Hone,	17
11.	Paraffin Soldering Wire,	18
12.	Cementing Hardened Tissue to Cork,	19
13.	Needle for Handling Sections, Covers, etc.,	20
14.	Diagram Illustrating Steps in Staining with Hæma.,	25
15.	Diagram Illustrating Steps in Staining with Hæma. and Eosin,	26
16.	Diagram Illustrating Steps in Staining with Borax-Carmine,	26
17.	Section Lifter,	32
18.	Appearance of Balsam Mounted Specimen,	33
19.	Mode of Handling Cover-glass,	32
20.	Diagram Showing Effect of Oil and Air Globules,	36
21.	Extraneous Substances—Hairs, etc.,	37
22.	Extraneous Substances—Starch, etc.,	38
23.	Elements of a Typical Cell,	39
24.	Structure of a Cell-Nucleus,	40
25.	Squamous Cells from Saliva,	41
26.	Pavement Epithelium,	42
27.	Frog's Mesentery—Silver Staining,	43
28.	Columnar Cells from Intestine,	46
29.	Ciliated Cells from Bronchus,	46
30.	Diagram Showing Organs of the Oyster,	46
31.	Corpuscular Elements of Human Blood,	47
32.	Diagram. Human Red-Blood Corpuscle in Profile,	48
33.	Blood Plaques,	49
34.	Glandular Cells from Liver,	50
35.	Fibrillated Connective Tissue,	52
36.	Yellow Elastic Tissue,	53
37.	Transverse Section of Ligamentum Nuchæ,	53

LIST OF ILLUSTRATIONS.

FIG.		PAGE
38.	Cells containing Fat,	55
39.	Adipose Tissue from Omentum,	55
40.	Hyaline Cartilage from Bronchus,	56
41.	Fibro-Cartilage from Intervertebral Disc,	57
42.	Elastic Cartilage from Pinna of Ox,	57
43.	Bone—Showing Laminated Structures,	58
44.	Bone—Showing Haversian Systems,	59
45.	Contents of Haversian Canals,	60
46.	Contents of Bone Lacuna,	60
47.	Bladder of Frog—Showing Non-Striated Muscle,	62
48.	Diagram—Illustrating Structure of Striated Muscular Fibre,	63
49.	Striated Muscular Fibre from Tongue,	64
50.	Cardiac Muscular Fibre,	65
51.	Artery in Transverse Section,	66
52.	Blood Capillaries,	67
53.	Layers of the Epidermis,	69
54.	Structure of the Derma,	70
55.	Hair in Transverse Section,	71
56.	Hair Follicle,	72
57.	Sudoriferous Gland,	73
58.	Sebaceous Gland,	73
59.	Skin in Vertical Section,	75
60.	Human Canine Tooth in Vertical Section,	79
61.	Fang of Tooth in Transverse Section,	82
62.	Stomach. Diagram,	83
63.	Cardiac Gastric Glands,	85
64.	Pyloric Gastric Glands,	86
65.	Stomach of Dog,	87
66.	Diagram Illustrating Intestinal Secretion,	89
67.	Diagram of Intestinal Absorption,	90
68.	Small Intestine with Peyer's Lymphatics,	92
69.	Bronchial Tube Arrangement,	94
70.	Bronchial Tube—Small,	96
71.	Bronchial Tube—Medium,	98
72.	Pulmonary Lobule—Perspective,	101
73.	Pulmonary Lobule—Longitudinal Section,	101
74.	Pulmonary Alveolus—Capillaries Filled,	102
75.	Lung of Pig,	103
76.	Pulmonary Alveolus Showing Lining,	104
77.	Liver. Diagram Illustrating Plan of Structure,	107
78.	Glandular Cells in Connection with Blood-vessels and Ducts,	109
79.	Liver of Pig,	111
80.	Human Liver—Low Power,	114
81.	Portal Canal,	115
82.	Hepatic Cells—Detached,	117
83.	Hepatic Lobule in Transverse Section,	118
84.	Bile Capillaries—Origin of Bile Duct,	119
85.	Kidney—Diagram Illustrating Plan of Structure,	121
86.	Kidney Tubules—Isolated,	123

LIST OF ILLUSTRATIONS.

FIG.		PAGE
87.	Blood-vessels—Arrangement in Kidney,	125
88.	Kidney—Low Power,	128
89.	Kidney—Cortex in Vertical Section,	130
90.	Kidney—Medulla in Longitudinal Section,	132
91.	Kidney—Medulla in Transverse Section,	133
92.	Uterus with Vaginal cul-de-sac,	136
93.	External Uterine Os,	138
94.	Vaginal Epithelium,	139
95.	Epithelium of Ureter,	140
96.	Epithelium of Urinary Bladder,	142
97.	Epithelium from Urinary Deposit,	143
98.	Ovary—Adult,	145
99.	Ovary—Fœtal,	148
100.	Suprarenal Capsule—Low Power,	151
101.	Suprarenal Capsule—High Power,	152
102.	Simple Tubular Gland,	153
103.	Coiled Tubular Gland,	154
104.	Branched Tubular Gland,	155
105.	Dilated Tubular Gland,	156
106.	Parotid Gland,	156
107.	Submaxillary Gland,	157
108.	Pancreas,	158
109.	Perivascular Lymph Spaces,	162
110.	Lymphatics of Central Tendon of the Diaphragm—Low Power,	165
111.	Lymphatics of Central Tendon of the Diaphragm—High Power,	166
112.	Lymph Node—Diagrammatic,	168
113.	Mesenteric Lymph Node—Low Power,	170
114.	Mesenteric Lymph Node—High Power,	171
115.	Blood-vessel Arrangement in the Spleen,	173
116.	Spleen,	175
117.	Thymus Body,	178
118.	Nerve Fibre,	180
119.	Connective Tissue of Nerve Trunk,	182
120.	Connective Tissue of Brain,	183
121.	Spinal Cord—Diagram,	185
122.	Cervical Spinal Cord—Transverse Section,	188
123.	Anterior Cornu of Gray Matter—Cervical Spinal Cord,	189
124.	Cerebrum,	193
125.	Cerebellum—Low Power,	194
126.	Cerebellum—High Power,	195

PRACTICAL MICROSCOPY.

PART FIRST.

TECHNOLOGY.

THE LABORATORY MICROSCOPE.

The histologist should be provided with a microscope, in which the principal features of the laboratory instrument, Fig. 1, are embraced.

The body A, which carries the optical parts, is made of two pieces of brass tubing, one sliding within the other and providing for alterations in length. The *objectives*, C, D, are attached to the body by means of the *angular carrier* E. The carrier is so pivoted that either objective may be turned into the optical axis, at will. The *eye-piece*, F, slips into the upper part of the body, with but little friction, so that it may be quickly and easily removed.

The *coarse or quick adjustment* for focussing consists of a rack G, which is attached to the body, and into this gears a small (concealed) pinion turned by the milled-head H.

The fine steel screw I, by means of which the more delicate adjustments for focussing are accomplished, terminates below in a hardened point, which impinges upon one end of a lever (concealed in the arm), the fulcrum of the same being indicated at the point J. The opposite end of the lever is inserted in a notch in the split arm K. By turning the milled-head L, the lever is moved, and the optical body raised or lowered with extreme delicacy.

The *stage*, upon which objects are placed for examination, is perforated at M, and a rotating disc—not indicated in the drawing —enables one to alter the size of the opening at will. Below the stage an arm may be seen which carries a fork supporting the mirror N.

The whole is supported on a short, stout pillar rising from the *foot* O.

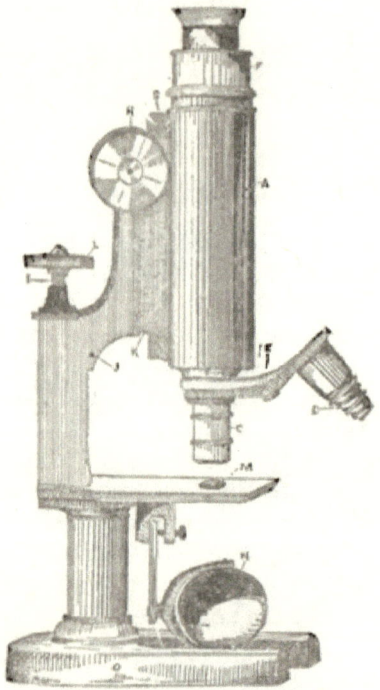

FIG. 1.—THE LABORATORY MICROSCOPE.

This instrument was designed and constructed for the laboratories of the New York University Medical College. It is strongly built; the mechanism is simple; and the height—10½ inches—not too great for use in the vertical position.

LENSES OF THE MICROSCOPE.

Fig. 2 shows the arrangement of lenses, including a high-power objective of the Wenham construction.

The *objective* A is provided with one simple and two compound lenses. The lens B, nearest the object, and the one upon which the magnifying power mainly depends, is an hemisphere of crown glass. Such a figured glass possesses both chromatic and spherical aberration in high degree. These faults are corrected by the compound flint and crown lenses, C and D, placed above the hemispherical glass.

The *eye-piece* consists of two crown-glass, plano-convex lenses, with their plane surface upward. The lower, E, is known as the field-lens, the upper, F, as the eye-lens. Eye-pieces add very materially to the magnifying power of the instrument, and are constructed of various strengths depending upon the curvature of the

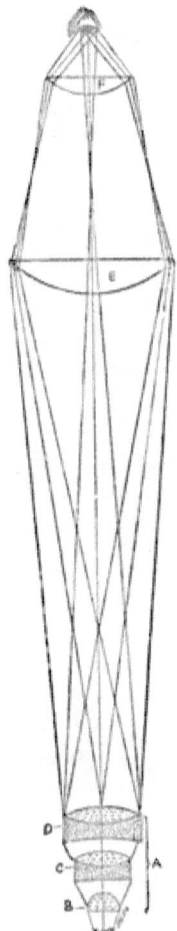

Fig. 2.—Diagram Showing the Relation of the Objective to the Eye-piece.

lenses. They are named according to power A, B, C, etc. The medium, B, is more commonly employed.

The microscope previously described stands, with the draw-tube in place, about ten and one-half inches high; and represents the instruments used in the New York University Laboratory of Biology and Pathology. They were constructed by Schrauer, of this city, costing about fifty-five dollars each. They are provided with a single eye-piece, and Hartnack objectives Nos. 2 and 7, giving from 30 to 400 diameters. Such an instrument is well adapted to the work of normal and pathological histology, though a condenser* should be attached below the stage and in the optical axis for high-power work with immersion lenses, and especially for bacteriological research. The stand is a rigid one, and if the height of the table upon which it is placed and the chair of the observer be in a proper relation, no discomfort need be experienced in using the microscope in the vertical position.

ADJUSTMENT OF THE MICROSCOPE.

The microscope should be placed in front of the observer, on a table of such height that, when seated, he may, by slightly inclining the head, and without bending the body, bring the eye easily over the eye-piece. The slightest straining of the body or neck should be avoided. The light should always be taken from the side, and it matters little which side. Clouds or clear sky serve as the best source of light for our present work. Always avoid direct sunlight. If artificial illumination be employed—though it is not advised for prolonged investigation—a small coal-oil flame may be tempered by blue glass.

ADJUSTMENT FOR ILLUMINATION.

It will be observed that there are two mirrors in the circular frame below the stage—one plane and the other concave. The latter will be employed almost exclusively in the work of this volume; and its curvature is such that parallel rays, impinging upon its surface, are focussed about two inches from the mirror. It will

---

* A non-achromatic condenser, after the formula of Abbé, of Jena, is in quite general use in this country. Its value has been very markedly increased here by the addition of a rack-and-pinion motion. In use for high-power work with tissues, it is first placed so that the plane surface of the upper lens is in contact with the under surface of the glass slip holding the object to be examined. Light is then reflected into the condenser as usual, excepting that the plane mirror is employed. This will give a strong illumination, but too diffuse for tissues. The light is then modified by diaphragms, or by racking the condenser downward until the best effect is secured. For bacterial search the strong illumination is employed. This gives prominence to the stained microbes, as the other elements in the field are lost in the excess of diffuse light.

also be noticed that the bar, carrying the mirror-fork, may be made to swing the mirror from side to side. The work which we are about to undertake is of such a character as to require the avoidance of oblique illumination. We must, therefore, keep our mirror-bar strictly in the vertical position. If—the mirror-bar being vertical—a line be drawn from the centre of the face of the mirror, through the opening (diaphragm) in the stage, passing on through the objective, and so continued upward through the body and the eye-piece, such a line would pass through the *optical axis*. The centre of the face of the mirror must be in this axis. If, then, having gotten the mirror-bar properly fixed once for all, the light from the adjacent right or left hand window impinges upon the concave surface of the mirror, and the latter be properly inclined, the rays will pass through the diaphragm in the stage, and become focussed a little above the same. The light rays will afterward diverge, enter the objective, and finally reach the eye of the observer.

The *field* of view (as the area seen in the microscope is termed) we will suppose to have been properly illuminated—and by this I mean that it presents as a clear, evenly-lighted area. Turn all the factors spoken of out of adjustment, and proceed to readjust. Observe that, if the mirror be turned—not swung—slightly out of proper position, one side of the field will appear dim or cloudy: this must be corrected, and the student must practise until this adjustment becomes easy of accomplishment. Then proceed to the

ADJUSTMENT FOR FOCUS.

Observe that the largest opening in the stage diaphragm is in the optical axis. Swing the low-power objective into use, and rack the tube up or down until it is about one inch from the stage.

Place a mounted object upon the stage (a stained section of some organ—say kidney—will be preferable). Examine the field through the eye-piece, and it will be found obscured by the stained object, and perhaps a dim notion of figure may be made out. Rack the body down carefully, watching the effect. The image becomes more and more distinct until, at a certain point, the best effect is secured. The object is *in focus*.

Note carefully the distance between the object and the objective (with the Hartnack No. 2 this will be about seven-eighths of an inch), and hereafter you will be able to focus more quickly.

Having observed the details of structure as shown with the inch objective, swing the high power into use. Rack the tube down, until the objective is about one-eighth of an inch from the glass covering the object. The field is much obscured. Watching the effect through the eye-piece, rack the tube down with great care until the image appears sharp. Note the distance with this objective as before with the low power, probably about one-thirty-second of an inch. Then endeavor, by slight alterations in the inclination of the mirror, to increase the illumination. Turn the diaphragm so that the light passes through a small opening, and note the improvement in definition. The rule is: *The higher the power, the smaller the diaphragm.*

You have doubtless observed, before this, that you cannot control the focussing as easily as when the low power was in use. Slight movements of the rack-work produce marked changes in definition; and it is difficult, with the coarse adjustment alone, to make as slight movements as you may desire. Recourse must be had to the fine adjustment.

Place the tip of the forefinger (either) upon the milled-head of the fine focussing-screw, and the ball of the thumb against its side, so that the hand is in an easy position. By a little lateral pressure the milled-head may be turned slightly either way. Note the effect on the image. You thus have the focussing under the most perfect control.

Remember that the fine adjustment is only necessary with high powers, and then only after the image has been found with the coarse adjustment.

METHOD IN OBSERVATION.

The study of objects under the microscope should be conducted with order or method.

The body being in the position before advised, so that the sitting may be prolonged without fatigue, let one hand be occupied in the maintenance of the focal adjustment. It will be found, however flat an object may seem to the unaided eye, that as it is moved so as to present different areas for examination (and with the higher powers only a small area can be seen at once), constant manipulation with the fine adjustment will be required. It will also be found that even the various parts of a simple histological element—like a cell—cannot be seen sharply with a single focal adjustment. The forefinger and thumb of one hand must be kept constantly on the milled-head of the fine focussing-screw. Supposing the light to be on our right, we devote the right hand to the focussing.

The left hand will be engaged with the glass slip upon which the object has been mounted. The forearm resting upon the table, let the thumb and forefinger rest on the left upper side of the stage, just touching the edges of the glass slip. The slightest pressure will then enable you to move the slip smoothly, steadily, and delicately.

Proceed to examine the object with method. Suppose a section of some tissue to be under examination—say one-fourth of an inch square. With the high power you will be able to see only a small fraction of the area at once. Commence at one corner to observe; and with the left hand move the object slowly in successive parallel lines, preserving the focus with the right hand, until the whole area of the section has been traversed.

Practice will soon establish perfect co-ordination of the movements involved, and will result in the ability to work with ease, celerity, and profit.

CONSERVATION OF THE EYESIGHT.

The beginner should not become accustomed to the use of one eye alone, or of closing either, in microscopical work. It will require but little practice to use the eyes alternately, and the retinal image of the unemployed eye will soon be ignored and unnoticed.

MAGNIFYING POWER AND MEASUREMENT OF OBJECTS.

The microscope is not, as the beginner usually supposes, to be valued according to its power of magnification, but rather according to the clearness and sharpness of the image afforded.

Magnifying power is generally expressed in diameters. A certain area is by the instrument made to appear, say, ten times as large as it appears to the naked eye. This object has, then, its apparent area increased one hundred times; but reference is made in describing such phenomena only to amplification in a single direction. The diameter has been increased ten times and would be expressed by prefixing the sign of multiplication, $e.g.$, $\times 10$.

A convenient unit of approximate measurement for the histologist is the apparent size of a human red blood-corpuscle with a given objective. Thirty-two hundred corpuscles, placed side by side, would measure one inch; or, we say, the diameter of a single corpuscle is the thirty-two-hundredth of an inch. After considerable practice, you will become accustomed to the apparent size of this object with a certain objective and eye-piece. This will aid in an approximate measurement of objects by comparison, and will further give the magnifying power of the microscope. If a corpuscle appears magnified to one inch in diameter, it is evident that the instrument magnifies thirty-two-hundred times. Should the diameter appear one-quarter of an inch, the power is eight hundred; one-eighth of an inch, four hundred, etc. The instrument which I have heretofore described, with the high power in use and the tube withdrawn, will present the corpuscle as averaging very nearly one-eighth of an inch in diameter—× 400. While this gives a gross idea of amplification, the method will often prove inaccurate because of individual errors in the estimation of proportions.

Use of the Stage-Micrometer.—From a dealer in optical goods purchase a Rogers* glass stage-micrometer, ruled in hundredths, thousandths, and five-thousandths of an inch. Also procure from the dealer in drawing instruments a two-inch boxwood rule divided decimally to fiftieths.

Place the micrometer on the stage of the microscope and focus the lines. Then place the rule also on the stage, but just in front of and parallel with the micrometer. By a little practice, using both eyes, the two rulings may be seen simultaneously, and by adjusting the position of the rule, the lines may be made to appear superposed.

Let us suppose that, with a given eye-piece and objective, the thousandth divisions on the micrometer correspond exactly with one of the tenths of the rule. Keeping this in mind, remove the micrometer scale and substitute an object, say a blood slide. Let us again suppose that the image of a given red corpuscle appears to cover three of the one-tenth-inch rulings, the latter scale having been left in position. It is evident that, as we found the value of one of the rule tenths to be, by the micrometer, the one-thousandth of an inch, the globule measures one-three-thousandth of an inch in diameter.

The value of the rule divisions must be determined for each objective; and a memorandum will then provide the means of quickly obtaining a very close approximation of the size of objects as viewed in the microscope, and at the same time indicate the degree of amplification of the instrument itself.

SKETCHING FROM THE MICROSCOPE.

Let me most emphatically urge the practice of sketching in connection with microscopy. "I am no artist," or "I have no skill

---

* The micrometer rulings of Professor Rogers, of Cambridge University, are without doubt of surpassing excellence. They are the result of many years of unwearying experimentation and are recognized standards throughout the scientific world.

in drawing," is often the reply to my advice in this matter. I then suggest that no special skill is needed to begin with, only patience and a dogged determination to succeed. The pictures in the microscopic field have no perspective, and may be reproduced in outline merely. Begin with simple tissues, reserving intricate detail until a short period of practice gives the technique needed. I do not recommend the camera lucida, as my experience strongly impresses me with this as a fact, that he who cannot sketch without a camera will never sketch with one. Pencil drawings may be very effectively colored with our staining fluids, diluted if necessary.

PREPARATIONS OF TISSUES FOR MICROSCOPICAL PURPOSES.

TISSUES ARE STUDIED BY TRANSMITTED LIGHT.

The microscopical study of both normal and pathological tissues is invariably conducted by the aid of transmitted light.

Tissues, if not naturally of sufficient delicacy to transmit light, must in some way be made translucent.

Delicate tissues like omenta, desquamating epithelia, fluids containing morphological elements, certain fibres, etc., are sufficiently diaphanous, and require no preparation. Such objects are simply placed upon the glass slip, a drop of some liquid added, and, when protected by a thin covering glass, are ready for the stage of the microscope.

PREPARATION BY TEASING.

The elements of structures mainly fibrous—*e.g.*, muscle, nerve, ligament, etc.—are well studied after a process of separation, by means of needles, known as teasing. A minute fragment of the organ or part having been isolated by the knife or scissors is placed upon a glass slip, and a drop of some fluid which will not alter the tissue added. Stout sewing-needles, stuck in slender wood handles, are commonly employed in the teasing process. The separation of tissues is frequently facilitated by means of dissociating fluids which remove the cement substance.

SECTION CUTTING.

After having become familiar with the various elementary structures of animal tissues, we proceed to the study of their relation to organs.

As the teasing process is not available with such complicated structures as lung, liver, kidney, brain, etc., we resort to methods of slicing—*i.e.*, *section cutting.*

Sections must be made of extreme tenuity, in order that the naturally opaque structures may be illuminated *by transmitted light.* This becomes an easy matter with such tissues as cartilage; but some, like bone, are much too hard to admit of cutting, and others are as much too soft; so that while certain tissues must be softened, the majority must be hardened. Fortunately, both of these conditions may be secured without in any way altering the appearance or relations of the structures. Hardening processes, from necessity, become a prominent feature in histological work; but I propose here to indicate some of the more useful methods of section cutting, reserving the hardening processes for another place.

FREE-HAND SECTION CUTTING.

My students, when ready for this work, are provided with some tissue which has been previously hardened. We will take, for ex-

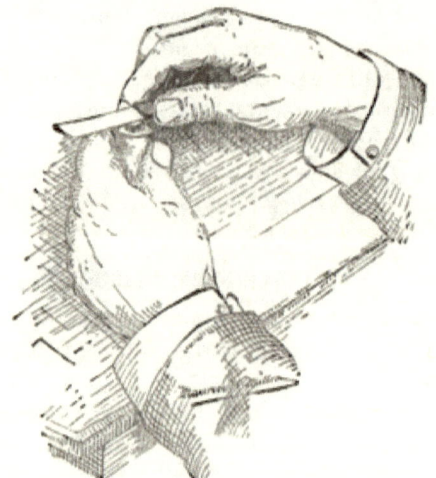

Fig. 3.—Free-Hand Section Cutting.

ample, a piece of liver which has been rendered sufficiently firm for our work by immersion in alcohol, and proceed to direct the steps in obtaining suitable sections by the simple free-hand method.

I wish to strongly emphasize the importance of this mode of cutting. A moderate amount of practice will render the microscopist independent of all appliances, save those of the most simple character and which are always obtainable.

An ordinary razor, with keen edge, and a shallow dish, preferably a saucer, partly filled with alcohol, are required. The razor best adapted to the work is concave on one side (the upper side as seen in Fig. 3) and nearly flat on the other, although this is largely a matter of personal preference.

Fig. 3 indicates the proper position of the hands in commencing the cut. I have made the sketch from a photograph of my esteemed colleague, Dr. Wesley M. Carpenter. The student should be seated at a table of such height as to afford a convenient rest for the forearms. A small piece of tissue is held between the thumb and forefinger of the left hand, so that it projects slightly above both. (In the cut, a cube of tissue, too small to handle in this way, has been cemented to a cork with paraffin in the manner hereafter described, and the cork held as just mentioned.) The hand carrying the tissue is held over the saucer of alcohol. The razor, held lightly in the right hand, as seen in the figure, is, previous to making every cut, dipped flatwise into the alcohol, so as to wet it thoroughly; and is then lifted horizontally, carrying several drops, perhaps half a drachm, of the fluid on the concave, upper surface. The alcohol serves to prevent the section from adhering to the knife, and to moisten the tissue. If allowed to become dry, the latter would be ruined by alterations of structure.

Now as to the manner of moving the knife. Resting the under surface upon the forefinger for steadiness, bring the edge of the blade nearest the heel to the margin of the tissue furthest from you. Then, entering the edge just below the upper surface of the tissue, with a light but steady hold draw the knife toward the right, at the same time advancing the edge toward the body. This passes the knife through the tissue diagonally, and leaves the upper surface of the latter perfectly flat or level. Remove the piece which has been cut and repeat the operation. Do not attempt to cut large or very thin sections at first. A minute fragment, if thin, is valuable.

As the razor is drawn through the tissue, the section floats in the alcohol; depress the point of the knife and the section will slide into the saucer of spirit, and thus prevent its injury. If it does not leave the knife readily, brush it along with a camel's-hair pencil which has been well wetted with the alcohol.

Proceed in the above manner until the tissue is exhausted, when you will have a great number of sections, large and small, thick and thin. Selecting the thinnest, lift them carefully with a needle, one at a time, into a small wide-mouthed bottle of alcohol; cork and label for future use.

When the work is finished, and before the spirit has evaporated from your fingers—it is impossible to avoid wetting the skin more or less—wash them thoroughly and wipe dry. This saves the roughening of the hands which is apt to result when alcohol has been allowed to dry upon them repeatedly.

SECTION CUTTING WITH THE STIRLING MICROTOME.

Of the numerous mechanical aids to section cutting, I shall mention only two or three. One of the earlier and better-known instruments is seen in Fig. 4. The Stirling microtome consists

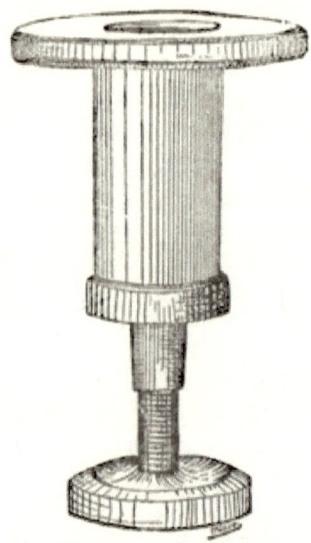

FIG. 4.—STIRLING'S MICROTOME.

essentially of a short brass tube, into which the tissue is fixed, either by pressure or by imbedding in wax. A screw enters below, which, acting on a plug, raises the contents of the tube. As the material to be cut is raised from time to time by the screw, it appears above the plate which surrounds the top of the tube. This plate steadies and guides the razor; and it is evident that more uniform sections may be cut with this little apparatus than would be possible with nothing to support the knife, or to regulate thickness, beyond the unaided skill of the operator.

Much depends upon the manner in which the material is fixed in the tube or well of the microtome. If the tissue be of a solid character, like liver, kidney, spleen, many tumors, etc., it may be surrounded with some carefully-fitted pieces of elder-pith,* carrot, etc., and the whole pressed evenly and quite firmly into the well. A small piece of tissue which, by cutting, can be made somewhat cubical in shape, may be surrounded by slabs of pith, carrot, or turnip, shaped as in Fig. 5. Indeed, the fragments of imbedding material can be shaped so as to fit tissue of almost any form. Before the whole is pressed into the well of the microtome, the bottom, against which the brass plug fits, should be cut off square.

The wax method of imbedding is employed with tissues such as brain, lung, soft tumors, etc., which might be injured by the previous treatment. To three parts of paraffin wax (a paraffin candle answers perfectly) add one part of vaselin, and heat until thoroughly mixed. The microtome having been previously warmed—standing upright—is filled with the imbedding mixture. The piece

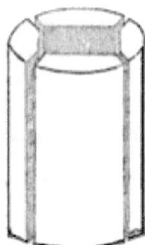

FIG. 5.—MANNER OF CUTTING AND ARRANGING PIECES OF PITH, TURNIP, ETC., FOR SUPPORTING HARDENED TISSUE IN THE WELL OF A MICROTOME.

of tissue is then carefully wiped dry with the blotting-paper and, just as the imbedding begins to congeal around the edges, is pressed below the surface with a needle and held until the cool mixture fixes it in position. The whole is now allowed to become thoroughly cold. By turning the screw, the plug of wax is raised; and it must be gradually cut away, by sliding the knife across the plate, until the upper part of the tissue appears.

Before commencing to cut sections—however the tissue may have been imbedded—provide yourself with a saucer of alcohol and a camel's-hair pencil. Having wetted the knife, turn the screw so that the tissue, with its imbedding, appears slightly above the plate of the microtome; and then, resting the blade of the razor on the

* The pith from the young shoots of *Ailantus glandulosus* (improperly called "Alanthus"), gathered in early autumn, is the best material for this method of imbedding with which I am acquainted. The wood is easily cut from the pith, and the latter is very large and firm.

plate (*vide* Fig. 6), make the cut precisely as in free-hand cutting. The section is then brushed off into the saucer, the screw turned up slightly, the razor wetted, and a second cut made. These steps are repeated until the required number of sections has been obtained.

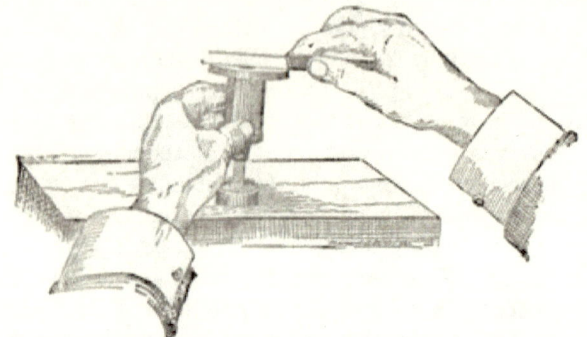

Fig. 6.—Method of Cutting Sections with the Stirling Microtome.

The imbedding will leave the cuts as they are floated in the alcohol. They may now be selected, lifted with the needle into clean spirit, and preserved as before indicated for future operations.

THE SCHRAUER MICROTOME.

Fig. 7 represents an improvement on the Stirling instrument, and a most convenient, practical, and inexpensive microtome for the physician.

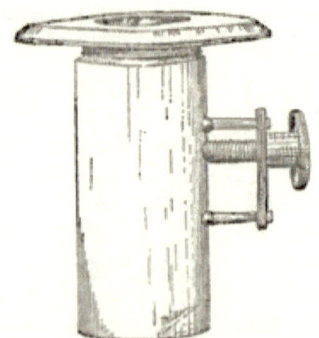

Fig. 7.—Schrauer's Improved Stirling Microtome.
With clamp for holding the tissue.

The tissue, if sufficiently hard, is held in a clamp or vice in the well of the instrument, the pressure being regulated by the side screw. By this means the necessity of imbedding is avoided. If the tissue be too soft to withstand such treatment, it is best cemented to a cork, and the cork then fastened in the clamp. A screw-head is cut upon a short cylinder, which works in a corresponding thread chased on the inside of the well-tube. The short cylinder carries the knife-plate, and, as the latter is turned to the right, the whole descends and the tissue projects, ready to be sliced off.

THE AUTHOR'S LABORATORY MICROTOME.

For certain work, some form of microtome becomes necessary in which the operator is relieved from supporting the knife. Fig. 8 is a sketch from such an instrument which I have contrived and which has been in daily use in my laboratory for over three years.

The carriage A, supporting the knife B, is of solid cast-iron;

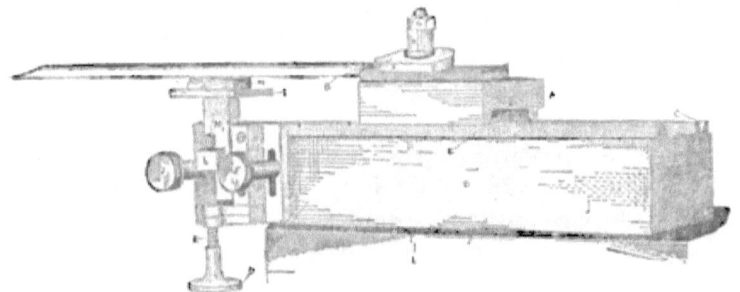

FIG. 8.—THE AUTHOR'S LABORATORY MICROTOME.

The instrument consists of a very heavy cast-iron bed upon which a carriage supporting a knife is made to slide. The tissue is cemented with paraffin (or held in an adjustable clamp not shown in the cut) to a table, which can be raised by a fine steel screw. The thickness of the section to be cut is controlled by turning the milled head actuating the finely threaded screw.

and has, upon the under side, a V guide, which fits into the longitudinal groove C of the base D. Parallel with this groove is a smooth, flat surface, upon which also travels the rib E of the knife-carriage. A second V has been avoided, in order to diminish friction. The knife is clamped rigidly to the upper surface of the carriage, by means of a Willis tool-holder consisting of a steel plate F, a nut G, and washer H.

The mechanism for supporting and positioning the tissue—not shown in the sketch—is built upon a plate I, which can be quickly fixed to the body of the microtome at the height and lateral incline required by the large set-screws J, J'. The mechanism for raising the tissue to the knife, between the cuts, consists of a screw K, of fifty threads to the inch, which, working in the nut L, elevates the bevelled slide M, to which the tissue N is affixed. An ether-freezing attachment may be substituted for the plate I.

The milled-head O is divided into one hundred parts, so that each fraction of a turn raises the tissue $\frac{1}{5000}$ of an inch.

The knife should possess an edge of the most exquisite keenness; and this holder admits the employment of almost any cutting instrument. In order to the production of the best results, the knife should be set at the most acute angle compatible with the use of the entire length of the cutting edge, from heel to point. Both knife and tissue are to be flooded with alcohol in ordinary work as in free-hand cutting. A drip-pan is provided and is placed below the tissue-carrier. A groove in the front upper edge of the base prevents the spirit from flowing over the track, which, mixing with the lubricating oil covering the latter, would interfere with the delicacy and ease of the sliding motion.

The value of this instrument is largely consequent upon its great solidity—the base weighing from eighteen to twenty-five pounds, with the knife-carriage correspondingly heavy. Just why such weight and solidity are necessary, and contribute so largely to our success in cutting sections, is not at once apparent. The microtome is now made by Mr. L. Schrauer, of New York, in two sizes; the smaller carrying a knife fourteen and the larger eighteen inches. A smaller pattern would present no special advantages over microtomes already in use.

SHARPENING KNIVES.

In the majority of instances of failure to produce suitable sections for microscopical work, the cause can be set down to dull knives; and I would urge the student to practise honing until able to put cutting instruments in good condition. If he will but start properly, success is sure. Nine-tenths of the microtomes are purchased because of failure in free-hand work with a dull knife; but *no advantage will be gained by a machine, if the student be incapable of keeping the knife up to a proper degree of keenness.*

A knife is a wedge, and for our purposes the edge must be of more than microscopical tenuity—it being impossible, with the microscope, to discover notches and nicks if properly sharpened.

It is impossible to secure the best results with indifferent tools. The knife should be hard enough to support an edge, but not so hard as to be brittle. The proper temper is about that given a good razor.

We need at least two hones—one comparatively coarse, for removing slight nicks, and another for finishing. The first part of the work is best done by means of a sort of artificial hone made with ground corundum. These are kept in stock by dealers in mechanics' supplies, of great variety in size and fineness. For razors a "00" corundum slip will best answer. This will very rapidly remove the inequalities from an exceedingly dull razor. A Turkish hone will be best for finishing. For my large knives I use a third, very soft and fine stone, known as *water-of-Ayr*.

Let the corundum slip be placed on a level support (mine are fitted into blocks like the carpenter's oil-stone), and cover the surface liberally with water.* The hones should always be worked

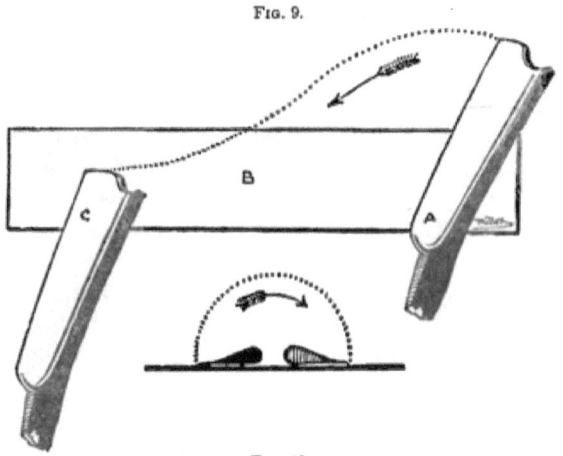

FIG. 9.

FIG. 10.

FIGS. 9 AND 10.—METHOD OF HONING.

The knife is first brought with its heel in the position shown at A, Fig. 9. It is then drawn forward as indicated by the curved dotted line until, at the end of the stroke, the position C is attained. Fig. 10 indicates the method of turning the blade before reversing and between each stroke.

wet. Place the knife flat on the stone near the right hand as at A, Fig. 9. Draw steadily in the direction of the curved dotted line—*i.e.*, from right to left—holding the blade firmly on the stone B with slight pressure until the position C is attained. Rotate the razor on its back—*vide* Fig. 10—so as to bring the other side on the stone; and draw from left to right. Observe that as the knife is drawn from side to side (the edge invariably looking toward the draw) it is always worked from *heel to point*. The amount of pressure may be proportioned to the condition of the edge. If it be

---

* A few drops of glycerin added to the water retards evaporation and appears to keep the surface of the hone in good condition.

2

badly nicked, considerable pressure may be employed; while, as it approaches keenness, the pressure is to be lessened, until the weight of the blade alone gives sufficient friction.

Repeat the process fifteen or twenty times, and examine the blade. If the nicks are yet visible, continue honing until they can no longer be seen. Then draw the edge across the thumb-nail. Do this lightly, and the sense of touch will reveal indentations which the eye failed to recognize. Continue the use of the coarse stone until the edge is perfect, as far as the thumb-nail test indicates.

The knife is then to be carefully wiped, so as to remove any coarse particles of corundum, and applied to the wetted Turkish hone with precisely the same motions as were employed in the first process. After a dozen or two strokes, examine the edge, by applying the palmar aspect of the thumb, with repeated light touches, from heel to point. This looks slightly dangerous to the novice, but it is an excellent method of determining the condition. Of course actual trial with a piece of hardened tissue is the best test.

Some most skilful technologists prefer to finish by stropping. I have not used a strop in my laboratory for over two years, preferring to use the knife as it comes, highly finished, from the water-of-Ayr hone. If a strop be employed, the leather should be glued smoothly to a support of wood, otherwise the edge of the knife will become rounded.

Stropping is conducted in the same manner as honing, only the edge of the knife *follows the stroke* instead of leading it.

SUPPORTING TISSUES FOR CUTTING.

Frequently small bits of tissue are required to be cut—pieces too small to be held with the fingers. I am in the habit of cement-

FIG. 11.—INSTRUMENT FOR SOLDERING TISSUE TO CORK SUPPORTS WITH PARAFFIN.
It consists of an awl handle of wood into which a short piece of wire, preferably copper, is driven and bent as shown.

ing such tissues into a hole in a bit of ailantus, or elder-pith, when the whole may be cut as one mass. Tissue is frequently cemented to cork for convenience of holding in free-hand cutting; or the cork may be held in the vise of the microtome. The edge of the knife should not be allowed to touch the cork.

Fig. 11 shows a simple little instrument, very convenient for using paraffin as a cement. A piece of stout copper or brass wire is bent as indicated, pointed, and driven into an ordinary awl-handle. Paraffin wax possesses the very valuable property of remaining solid at ordinary temperatures, not cracking in the cold of winter or softening in summer. It is unaltered by most reagents, is easily rendered fluid, and quickly solidifies. As a cement, it is invaluable to the microscopical technologist.

Fig. 12 indicates the method of cementing a piece of tissue to a

FIG. 12.—MODE OF CEMENTING TISSUE TO A CORK SUPPORT WITH PARAFFIN.

cork or other support. The tissue having been properly placed, the wire tool is heated for a moment in the alcohol flame, and then touched to a cake of paraffin. The paraffin is melted in the vicinity of the hot wire, a drop adheres to the latter and is carried to the edge of the tissue. In the cut the wire tool is seen in the position necessary for cementing one edge. The wire being removed, the wax immediately cools and becomes solid. The other sides are afterward cemented in like manner. The whole is done in less time than is necessary to the description of the process.

PREPARATION OF TISSUES FOR CUTTING, ETC.

We have already seen that most animal tissues are unsuitable for the production of thin sections until hardened.

It is also a fact, paradoxical though it may seem, that fresh tissues do not present truthful appearances of structural elements. The old-school histologists insisted upon the presentation of structures unaltered by chemical substances, while the modern worker has discarded such tissue with very few exceptions. Many descriptions for structure and growth, the result of study upon fresh material, have been proven by later methods grossly inaccurate.

It is impossible to remove tissues from the living animal and to subject them to microscopical observation without, at the same time, exposing them to such radical changes of environment as to produce structural alterations. Certain tissues, presenting in the living condition stellate cells with the most delicate, though well-defined branching processes, when removed from contact with the body, however expeditiously, afford no hint of anything resembling such elements, as they are quickly reduced to simple spherical outlines.

In short, it is impossible to study fresh material, as such, without constant danger of erroneous conclusions, as retrograde alterations of structure commence with surprising rapidity the moment a part is severed from the influences which control the complete organism.

From what has been said we appreciate the necessity of agents which, when applied to portions freshly removed from the animal, or even before removal, shall instantly stop all physiological processes and retain the elements in permanent fixity.

Very much of the human structure which is available will be secured only after functional activities have long ceased, and the structure essentially altered. We are, therefore, compelled to resort to the use of material from the lower animals in very many instances.

ALCOHOL HARDENING.

The tissue, whatever process may be in contemplation, having been removed from the body as quickly after death as possible, *without washing* or allowing contact with water in any way, should, with a sharp scalpel, be divided into small pieces. Of the more solid organs, pieces one-half inch square by one-fourth inch thick will be sufficiently small, and they will harden rapidly The smaller the pieces and the larger the quantity of hardening fluid, the more quickly will the process be completed. The volume of fluid should exceed that of the tissue at least twenty times. Wide-mouth, well-stoppered bottles, from one ounce to a pint, or even larger, are best; and they should be carefully labelled and kept in a cool place with occasional agitation.

Quick Method.—A piece of any solid organ, say liver, spleen, pancreas, kidney, uterus, lymph-node, etc., not larger than one-half inch square by one-eighth thick, may be perfectly hardened in twelve hours by immersion in one ounce of ninety-five per cent alcohol. No more should be thus prepared than is to be cut within twenty-four hours, on account of the shrinkage which results after the prolonged immersion of solid structures in strong spirit.

After the tissue has been one hour in the above, it may be hardened in one or two hours more, if transferred to absolute alcohol. This method is of frequent advantage in pathological histology.

Ordinary Method.—The method quite general here, and intended to prevent shrinkage, is as follows:

The organs, cut into pieces from one-half to three-fourths of an inch cube, are placed in a mixture of alcohol one part, water two parts (called in this laboratory "Alcohol A") for twelve hours. This removes the blood, and prepares the tissue for the next mixture—alcohol one part, water one part ("Alcohol B"), where it remains twenty-four hours. The pieces are afterward removed to ninety-five per cent alcohol ("Alcohol C"). The strong alcohol completes the hardening, and serves as a preservative until such time as sections may be required. The process is complete in from two to four weeks, and the material will keep without deterioration for three or four years, especially if the spirit be changed occasionally.

Ordinary anatomical specimens which have been preserved in dilute alcohol are of no value for our purpose.

CHROMIC-ACID FIXING AND HARDENING.

Chromic acid is a very deliquescent salt, and is best preserved by making a strong solution at once, and then diluting it as may be needed. A stock solution may be made as follows:

Chromic acid (crystals), . . 25 grammes.
Water (distilled or rain), . . 75 c.cm. M.

For general use dilute 20 parts with 600 parts of water, which gives a strength of nearly one-sixth of one per cent.

The tissue, as soon as secured and properly divided, is placed in the above, remembering the rule regarding quantity. Change in twenty-four hours to fresh solution, and again on the third day.

In seven days, or thereabout, change the fluid again. The tissue must now be watched carefully, and when, on cutting through a piece, the fluid is found to have stained the blocks completely, taking from two to three, or even four weeks, remove to a large jar of clear water and wash, changing the water frequently for twenty-four hours. The washing having removed the chromic acid, the tissue is further hardened in Alcohol A, B, C.

The special applications of this method, as well as of those which will follow, are indicated in Part Third.

MÜLLER'S FLUID (MODIFIED).*

Bichromate of potash,	25 grammes.
Sulphate of copper,	5 "
Water,	1,000 c.cm. M.

This may be employed in precisely the same manner as the dilute chromic-acid solution.

DECALCIFYING PROCESS.

6% chromic acid solution,	9 parts.
Nitric acid, C. P.,	1 part.
Water,	90 parts.

The earthy salts may be removed from teeth and small pieces of bone with a liberal supply of the above in about twenty days. A frequent change of the solution will greatly facilitate the process; and an occasional addition of a few drops of the nitric acid may be made, with very dense bone. After the removal of the lime salts, the pieces may be preserved in alcohol until such time as sections are needed, when they may be cut with the microtome without injury to the knife.

DISSOCIATING PROCESS (W. STIRLING).

Artificial Gastric Fluid.

Pepsin,	1 gramme.
Hydrochloric acid,	1 c.cm.
Water,	500 c.cm. M.

This process depends for its value upon the fact that certain connective tissues are more rapidly dissolved by the fluid than others.

* The original Müller's fluid consists of the above (minus the copper salt) with an addition of 12.5 grammes of sulphate of soda.

BAYBERRY TALLOW, HARDENING OR INFILTRATING PROCESS.

Some three years since, I devised a method of infiltrating tissues with bayberry tallow. Tissues like lung, etc., which are delicately cellular and hence very difficult to cut, when infiltrated with this material are supported in such a manner as to render the production of thin sections a very easy matter.

Bayberry tallow is found in commerce in various grades. The best is white, clean, and of a consistency about equal to that of hard mutton tallow. It is instantly soluble in benzol, and dissolves rather slowly in alcohol.

Having selected a piece of alcohol-hardened tissue for cutting, carefully wipe it dry with blotting-paper and drop it into a capsule containing melted bayberry tallow. In order to render the tallow sufficiently fluid, and yet prevent the heat from becoming great enough to injure the tissue, the capsule should be set over a water-bath. Bubbles immediately arise as the spirit is vaporized and the tallow gradually fills the interstices of the tissue. If the latter be of a somewhat dense character it will be best, before placing it in the tallow, to allow it to remain for an hour in pure benzol, which, evaporating at a very low temperature, gives more ready admission to the infiltrating medium.

The length of time required for complete infiltration will depend upon the density and the degree of heat employed. Usually from ten to thirty minutes will suffice.

The tissue, having become sufficiently infiltrated, is lifted out with the forceps, placed on a cork support, and allowed to cool. It is then cut, either free-handed or with the microtome, and *without alcohol*. The dry sections, resembling tallow or wax shavings, are brushed into a saucer of pure benzol, when in a moment the tallow will be dissolved from the tissue. The sections are then lifted with a needle singly into a saucer of alcohol to remove the benzol. Afterward they are transferred to a bottle of spirit, and labelled for future use. They will keep indefinitely.

This process is peculiarly advantageous with such tissues as lung, pancreas, cerebellum, intestine, etc., where the structures require support only while they are being cut. The infiltrated blocks of tissue can be kept dry until such time as they may be wanted.

CELLOIDIN INFILTRATION.

Certain structures require permanent support—*i.e.*, not only while being cut, but during the subsequent handling of the sections. The celloidin infiltrating process is best adapted to such material. Considerable time is needed for the successful employment of the process, but results can be secured that cannot be equalled with any other method.

Celloidin is the proprietary name of a sort of pyroxylin, very soluble in a mixture of ether and alcohol, producing a *collodion*. If thick collodion be exposed for a few moments to the air it becomes semi-solid—not unlike boiled egg-albumen; and to this property is due the value of a solution of celloidin in histology. It may be used as follows:

To a mixture of equal parts of ether and alcohol add celloidin * until the thickest possible solution has been obtained.

A piece of alcohol-hardened tissue having been selected and kept for the preceding twenty-four hours in a mixture of equal parts of alcohol and ether, is placed in about an ounce of the solution, and allowed to remain twenty-four hours. The bottle containing the whole should be well corked to prevent evaporation.

The tissue after infiltration is to be placed on a cork support and allowed to remain in the open air for a few minutes, after which it should be plunged into a mixture of alcohol two parts, water one part. Here it may remain for twenty-four hours, or until wanted.

Cut in the usual way, using a mixture of alcohol two parts, water one part, for flooding the knife; the section should be finally preserved in the same instead of pure alcohol, which would dissolve the celloidin.

In infiltrating the tissue with the collodion it is best, especially if it be very dense in parts, to use, first, a thin and subsequently the thick solution. A more perfect infiltration is often obtained in this way. In some cases I have been obliged to continue the maceration for several days. The solution should be kept in well-stoppered bottles, as the ether is exceedingly volatile. Should the collodion at any time become solid from evaporation, it may be easily dissolved by adding the ether and alcohol mixture.

The process is of inestimable value where delicate parts are weakly supported, and where it is important to preserve the normal relations. The gelatin-like collodion permeates every space, and as it is not to be removed in the future handling of the sections, it affords a support to portions that would otherwise be lost or distorted. It offers no obstruction to the light, being perfectly translucent and nearly colorless.

* I find, after repeated trial, that the ordinary soluble gun-cotton, such as is employed by photographers, is in no way inferior to the celloidin.

HARDENING BY FREEZING, ETC.

I do not recommend the freezing process.

Other fixing and hardening methods, which are of special application only, will be introduced in our future work as occasion may demand.

STAINING AGENTS AND METHODS.

STAINING FLUIDS.

It is a very interesting fact (and one upon which our present knowledge of histology largely depends) that, on examination of tissues which have been dyed with special colored fluids, the dye will be found to have colored certain anatomical elements very deeply, others slightly, while others still remain unstained. Not only are different depths of color thus obtained, but different tints, even with a single dye, are often presented. If a section of some animal tissue be immersed in a mordanted solution of logwood, for example, besides the different depths of blue which are communicated to certain parts, other elements present pink and violet tints in various shades.

The rule concerning the selection of dyes seems to be that those elements of a tissue which are the most highly endowed physiologically take the staining most readily. The minute granules of nuclei are so deeply stained in the logwood dye as to appear almost black. The nuclei are plainly stained, while the limiting membrane of cells is usually but slightly colored. Old, dense connective tissues stain feebly, or fail entirely to take color. The differentiation is, without doubt, due to chemical action between the elements of the dye and those of the tissue.

A very great number and variety of materials have been used for histological differentiation, and a simple enumeration of them all would very nearly fill the remainder of our pages. It will be found, however, that leading histologists confine themselves to two or three standard formulæ for general work. I shall notice only those methods which have been thoroughly demonstrated by years of employment as best for the purpose suggested. Special cases will require special treatment, which will be indicated in proper connection.

HÆMATOXYLIN* STAINING FLUID.

To about eight fluid-ounces of a hot, saturated aqueous solution of common *alum*, contained in a porcelain caspule, add, a few grains

---

* The coloring principle of the hæmatoxylon Campechianum. Merck's preparation should be used.

at a time, one drachm of *hæmatoxylin*, with constant stirring. Boil over the spirit-lamp very slowly for fifteen minutes. Add sufficient water to compensate for evaporation; and, when cold, pour into a wide-mouth bottle. Throw in a piece of camphor, say 30 grains, allow the whole to remain exposed to the air for one week, and then filter.

The solution should always be filtered before using. Keep the filter paper in a funnel, and use it as a stopper for the bottle. The dye improves in strength of color for two or three weeks.

Should the solution, which is of a beautiful purple or violet color, at any time turn red, a small piece of common chalk may be added. This will restore the color by neutralizing the acidity. A few crystals of alum should always be kept in the bottle to insure saturation.

Prepared as above, the dye will keep perfectly for at least eight months, and gives a permanent stain.

BORAX-CARMINE STAINING FLUID.

To eight ounces of a saturated aqueous solution of *borax* (borate of soda) add one drachm of the best No. 40 *carmine* (previously rubbed into a paste with a little water). In order to insure saturation, some borax crystals should always be left undissolved at the bottom of the bottle. Agitate frequently, and, after twenty-four hours, add fifteen drops of *liquor potassæ*.

Always filter or decant before using. It will keep indefinitely, improving, to a certain extent, with age.

EOSIN SOLUTION.

Alcohol,	2 ounces.
Eosin,	1 drachm.

This will give a saturated solution.

PICRIC-ACID SOLUTION.

Picric acid,	½ ounce.
Water,	4 ounces.

The acid is in excess, insuring saturation.

NITRATE-OF-SILVER SOLUTION.

Nitrate of silver,	5 grains.
Distilled water,	4 ounces.

If the water be pure, light will have no effect upon the solution.

STAINING METHODS.

HÆMATOXYLIN STAINING PROCESS.

You will require for future work a needle like Fig. 13, several saucers, preferably of white ware; a few watch-glasses—large, odd sizes are usually cheaply obtainable at the jeweller's; half a dozen glass salt-cellars—small ones known as "individual salts;" and a two-ounce, shallow, covered porcelain box, such as druggists use for ointments, dentifrices, etc.

Place on the work-table (best located so as to be lighted from your side and not from the front) in order, as in Fig. 14.

Fig. 13.—Needle for Lifting Sections, etc.

1. A *watch-glass*, containing say fl. ℨ ij. of hæmatoxylin fluid.
2. *Saucer*, filled with water.
3. *Salt-cellar*, filled with alcohol.
4. The *covered porcelain box*, containing about an ounce of oil of cloves.*

Select a section from some one of your stock bottles, lifting it out with the needle, and place it in the hæma. solution. The section having been taken from alcohol and transferred to an aqueous staining fluid, will twirl about on the surface of the latter, inasmuch as currents are formed by the union of the water and the spirit.

"How long shall I let the section remain in the hæma.?" The only answer I can give is, "Until properly stained." Nothing but experience will give you any more definite information. Much depends upon some peculiar property in the tissues: some stain rapidly, others stain very slowly. The strength of the dye is another determining factor. Usually with the hæma. formula, as given, from six to ten minutes will suffice.

Place the needle under the section (if the fluid be so opaque as to hide the tissue, place the watch-glass over a piece of white paper or a bit of mirror) and gently lift it out; drain off the adhering drop of dye on the edge of the glass, and drop into the saucer of water. Here we can judge as to the color, and we, perhaps, find it

---

* The oil of bergamot must be used for clarifying sections which have been infiltrated with collodion, as the clove oil is a solvent of the pyroxyline.

to be of a light purple—too light; so you may return it to the hæma. for another period of two or three minutes, which will probably give sufficient depth.

As the section floats on the washing water, you will notice that the latter will be colored by the dye, some of which leaves the tissue. Allow the water to act until no more color comes out. The tint of the section changes from purple to violet, and the water must be allowed to act until the change is complete. Again, you will remember that this dye contains alum, and if you hurry the washing you will undoubtedly find crystals covering your specimen after it has been mounted. From five to ten minutes will complete the washing.

If you were to examine your section at this stage, you would find it opaque, and as we are obliged to study our objects mainly by transmitted light, we must find some means of securing translucency. The essential oils are used for this purpose, oil of cloves being commonly employed. Lift the section from the water with the needle; let it drain a moment, and then drop it into the alcohol with which the salt-cellar was filled. The object of this bath is the removal of the water from the tissue, and this will be accomplished in from five to ten minutes. Again lift the section and place it in the oil of cloves. The tissue floats out flat, and in a few minutes sinks in the oil.

We might proceed to the examination of the stained section; but I shall ask you to let it remain in the oil, covering the box carefully to exclude our great enemy, the dust, until we have learned more about staining.

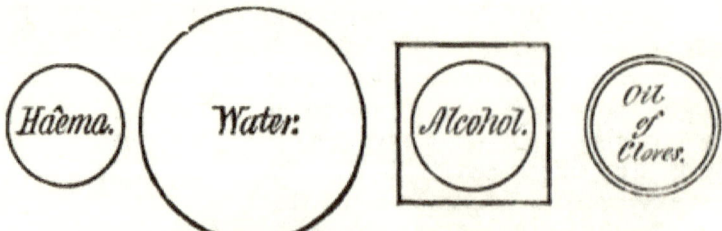

FIG. 14.—DIAGRAM INDICATING THE SUCCESSIVE STEPS IN STAINING WITH THE HÆMATOXYLIN SOLUTION.

To recapitulate: The essential steps in the hæma. process are:
1. Staining the tissue—hæma.
2. Washing—water.
3. Dehydrating—alcohol.
4. Rendering translucent—oil of cloves.

As the section is put in the dye, care should be taken to so float it out that it may not be curled. This is easily done with the needle. After the alcohol bath, however, this becomes difficult, as the tissue is rendered stiff by the removal of the water.

This is the simplest and best of all methods for general work, and you are advised to master every detail of the process. After reading the directions which I have given, and having never seen the work actually done, you will not be singular if you conclude the staining of tissues to be a tedious and slow process; but after a month's work you will be able to stain fifty different sections in half an hour, and have them ready for mounting.

HÆMATOXYLIN AND EOSIN. DOUBLE STAINING.

Very beautiful and valuable results in differentiation are obtained by staining first with hæma. and subsequently with eosin. Eosin is a salt of resorcin, staining most animal tissues pink, and it affords with the hæma. a good contrasting color. The tissue is to be stained in hæma. and washed in water as usual; then it is placed in the eosin solution, and afterward washed again. The subsequent treatment is as with the plain hæma. process, viz., dehydration with alcohol, after which the oil of cloves.

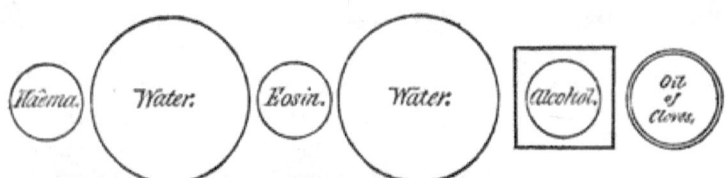

FIG. 15.—DIAGRAM INDICATING THE SUCCESSIVE STEPS IN DOUBLE STAINING WITH HÆMA. AND EOSIN.

The diagram, Fig. 15, shows the process complete:
1. *Watch-glass* with hæma.
2. *Saucer* with water.
3. *Watch-glass* two-thirds filled with water, with five drops of eosin solution added.
4. *Saucer* containing water.
5. *Salt-cellar* filled with alcohol.
6. *Covered oil-dish*.

The eosin stains very quickly, generally in about a minute. Care should be taken not to overstain with it, as it cannot be washed out. If the sections are found at any time to be overstained with hæma. the color may be removed to any desired extent by floating them in a *saturated aqueous solution of alum*. They must afterward be washed in clean water.

BORAX-CARMINE STAINING PROCESS.

Arrange your materials as in the diagram, Fig. 16.
1. *Watch-glass* two-thirds filled with the carmine fluid.
2. *Saucer* containing about an ounce of alcohol.
3. *Salt-cellar* filled with a saturated solution of oxalic acid in alcohol.
4. *Salt-cellar* with alcohol.
5. *Porcelain dish* containing oil of cloves.

The carmine solution will stain ordinarily in from three to ten minutes. After the section has been for a few minutes in the dye,

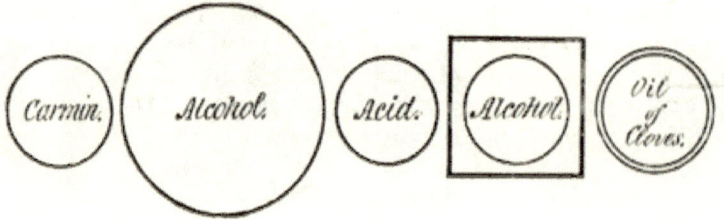

FIG. 16.—DIAGRAM INDICATING THE SUCCESSIVE STEPS IN STAINING WITH BORAX-CARMINE.

you will lift it with the needle, drain, and transfer to the saucer containing alcohol. You will then be enabled to determine whether the section is sufficiently stained; it should be a deep, opaque red. The alcohol washes off the section, removing the adhering solution of carmine.

The carmine must now be fixed in the tissue or *mordanted;* and this you proceed to do by transferring the section to the watch-glass of oxalic-acid solution. Notice the change in color, from a dull red to a bright crimson, and when the change is complete, lift it into the salt-cellar containing clean alcohol. This dissolves out the acid, which, if left, would appear later on the specimen in crystals. Five minutes suffice for this washing, after which transfer to the oil of cloves.

This process does not give as sharp contrasts as the hæma. and eosin, but it is simpler and very permanent. It is best to select some one process for general work, *and adhere to it.* The acid of the carmine process must be guarded with extreme care, as the smallest particle is sufficient to spoil the hæma. solution. Look to it that the dishes are kept scrupulously clean, and the same care must be bestowed upon the needles, forceps, etc.

You may, of course, stain several sections at once, providing you take care to keep them from rolling up or sticking together.

While the vessels which I have recommended will be found of convenient, proportionate, and economical size for general work, larger ones are sometimes needed; and almost any glass or porcelain vessel may be impressed for duty.

CARMINE AND PICRIC ACID STAINING.

After having washed the tissue, subsequently to mordanting with oxalic acid in the borax-carmine process, a bright yellow may be communicated to certain anatomical elements by means of picric acid. This often gives a valuable differentiation.

The sections are placed in the picric-acid solution and allowed to remain for ten minutes. Remove to water one ounce, glacial acetic acid ten drops for a moment, to fix the yellow; after this dehydrate with alcohol, and clarify with oil of cloves as usual. The sections should be transferred to the picric and acetic acid solutions by means of a platinum wire or a minute glass rod. The ordinary needle would be corroded, and the sections thereby discolored.

MOUNTING OBJECTS.

CLEANING SLIDES AND COVERS.

When purchasing slides, let me urge you to get them of good quality. The regular size is one by three inches, and the edges should be smoothed. As furnished by the dealers they are usually quite clean, and only require rubbing with a piece of old linen to prepare them for use.

The cover-glasses should be thin, not over $\frac{1}{100}$th of an inch, called in the trade-lists "No. 1." Circles or squares three-quarters of an inch in diameter are generally convenient. They must be thoroughly cleaned: Drop them singly into a saucer containing hydrochloric acid. Then pour off the acid, and let clean water run into the dish for several minutes. Drain off the water and pour an ounce of alcohol on the covers. Remove them one at a time with the forceps or needle, and wipe dry with old linen.* The

---

* We are indebted to Professor Gage, of Cornell University, for suggesting the use of Japanese tissue paper for wiping cover-glasses, lenses, etc. Ordinary manilla toilet paper is also an excellent material for such work.

glass may be held between the thumb and forefinger, the linen being interposed. Very slight pressure and rubbing will complete the process. The surface of the glasses should be brilliant, and they are to be preserved for future use in a dust-tight box.

TRANSFERRING THE SECTIONS TO THE SLIDE.

Procure a piece of either very thin sheet copper or heavy tin foil, three inches long and one-half inch wide, and bend it about three-fourths of an inch from one end, making a section lifter as shown in Fig. 17.

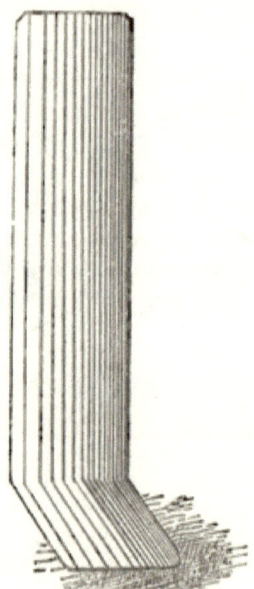

FIG. 17.—SECTION SPOON.
Strip of copper or heavy tin foil, best for lifting sections from staining and other fluids. For use in fluids which would attack metals, the spoon should be constructed from horn.

Place a clean slide on the table before you and with the section lifter used like a spoon dip up one of the sections from the clove oil. By inclining the lifter, the section may be made to float to the centre of the slide. A small sable brush is often convenient for coaxing the section off the lifter.

If it were our present object to simply examine the section, we could drop a thin cover-glass on the specimen, and it would be ready for study. Such an object would afford every requirement for present observation, but would not be permanent. The oil of cloves would evaporate after a few days and the section be ruined. We proceed to make a permanenet mounting of our object.

The clove oil, surrounding the section on the slide, is first to be removed; and it can easily be done by means of blotting-paper. With a narrow slip of thin filter paper wipe up the oil, exercising care not to touch the section or it will become torn. Proceed carefully, taking fresh paper until the oil will no longer drain from the

FIG. 18.—METHOD OF LABELLING A MOUNTED SPECIMEN.

section when the slide is held vertically. With a glass rod remove a little of the dammar solution (*vide* formulæ) from the bottle and allow a drop of this varnish to fall upon the section.

Pick up a clean cover-glass with a needle, and place it on the drop of dammar. This operation is seen in Fig. 18. The point of the needle may be placed beneath the cover-glass, the tip of the

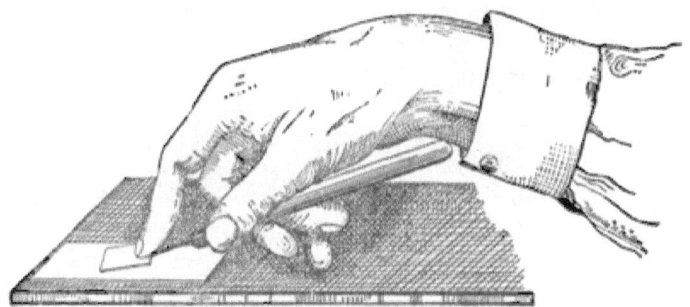

FIG. 19.—MODE OF HANDLING THE COVER-GLASS IN MOUNTING TISSUES.

forefinger pressing lightly over it, and you will be enabled to carry the thin glass wherever desired.

As the cover settles down the air is pressed out, until finally the section appears imbedded in the varnish—the latter filling the space between the cover and the slide.

The object is "mounted." You have a *permanent specimen*. The slide must be kept flat, as the dammar is soft. After some weeks, the varnish around the edge of the cover will stiffen, and eventually become solid. Do not paint colored rings around the specimen. Nothing can present a neater appearance than the simple mount, as I have described it, after having been properly labelled. Labels seven-eighths of an inch square may be put on one or both ends, with the name of the object, date, method of staining, or whatever particulars you may prefer.

Specimens should be kept in trays or boxes so as to always lie flat.

CARE OF THE MICROSCOPE.

The objectives constitute the most valuable part of the instrument. The lenses should never be touched with the fingers; indeed the same rule applies to all optical surfaces. When the glasses become soiled, they may be cleaned, but it should be done with great care. While the effect of a single cleaning would probably not be to the slightest appreciable injury to the glass, repeated wiping with any material, however soft, will destroy the perfect polish, and result in obstruction of light and consequent dimness in the field. Never use a chamois leather on an optical surface, as these skins contain gritty particles. Old, well-worn linen and Japanese paper are by far the best materials for wiping glasses. If a lens be covered with dust, brush it off; breathe on the surface, and wipe gently with the linen or paper. Should you get clove oil on the front lens of the objective (as frequently happens when examining temporary mounts) wipe it dry and then clean with the linen moistened with a drop of alcohol. Dammar varnish can be very readily removed from any surface after having softened it with oil of cloves. The front lens of the objective, being the only one exposed, is the one usually soiled.

Particles of dirt on the objective, as I have said, cause a dimness in the field—the image is blurred. Dust on the lenses of the eye-piece, however, appears in the field. These lenses are readily cleaned by dusting, and wiping with the linen, after having breathed on the surface. Never wipe a lens when dusting with a camel's-hair brush will answer the purpose.

The microscope should either be covered with a shade or cloth, or put away in its case, when not in use. The delicate mechanism of the fine adjustment becomes worn and shaky if not kept free from dirt.

PART SECOND.

STRUCTURAL ELEMENTS.

PRELIMINARY STUDY.

FORM OF OBJECTS.

From a single and hasty view of bodies under the microscope, we are liable to form erroneous ideas of form. Either a sphere, disc, ellipsoid, ovoid, or cone may be so viewed as to present a circular outline. It therefore becomes important to view objects in more than a single position. This can easily be accomplished with isolated particles by suspension in a liquid. In this way the true shape of a blood-corpuscle, *e.g.*, may be determined.

Again, much information concerning the actual form of bodies may be gained by a proper adjustment of the fine focussing screw. You may remember that the depth of the field of view in the microscope is exceedingly slight. Speaking accurately, only a single plane can be seen with a single focal adjustment; but by gradually raising or lowering the tube of the microscope, the different parts of a body may be focussed and studied and an accurate idea of form secured.

With a glass rod place a drop of milk, which has previously been diluted with three parts of water, on a slide, and put a cover-glass thereon as in Fig. 19. Focus first with the low power (L). A multitude of minute dots are observed. Then switch on the high power (H), and the dots will resolve into circular figures. Select one of the smaller particles and, as you raise the focus, the centre of the figure retains its brilliancy, while the edges become dark or blurred, showing convexity. Reverse the focus, and the centre again retains its sharpness long after the edge has become blurred. The figure, then, is a spheroid. These bodies are fat-globules. Particles of free fat always assume the spheroidal form when suspended in a liquid.

Note the larger globules; they have become flattened by pressure of the cover-glass.

Clean the slide, and make a second preparation from the diluted milk—first, however, shaking it violently in a bottle. Note the flattened air-bubbles among the oil-globules. Observe that these air-bubbles have no intrinsic color, while the fat-globules are faintly

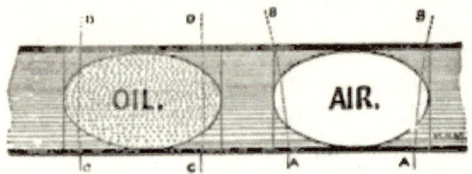

FIG. 20.—DIAGRAM SHOWING THE EFFECT OF AIR-BUBBLES AND OIL-GLOBULES IN A MOUNTED SPECIMEN UPON THE RAYS OF LIGHT.

The lines A, B show the refraction of the rays (so as to produce a ring of color) by the action of two plano-concave water lenses which are formed by the air-bubble.

The oil is seen to correct the refraction of C D, thus giving but little color to the margin of this globule.

yellow. Observe the change in the ring of prismatic color about the edge of the air-bubble, as the focus is altered. No such color will be seen in connection with the oil-globule.

The bubbles assume various figures from pressure of the cover-glass.

MOVEMENT OF OBJECTS.

Objects are frequently seen moving in the field of the microscope, the movement being magnified equally with their dimensions.

Thermal Currents.—When with the previous specimen, or any other fluid mount, the warm hand is brought close to one side of the stage, the globules in the field will be seen swimming more or less rapidly. These currents are due to the unequal heating of the liquid under observation. The direction of the current is in the reverse of its apparent motion.

Brownian Movement.—Place a fragment of dry carmine on a slide; add a drop of water, and with a needle stir until a paste is formed. Add another drop of water, and immediately put on the cover-glass. With H, note the most minute particles, and observe their peculiar, dancing motion. This occurs when almost any finely-divided and generally insoluble solid is mixed with water; it ceases after a short time. The movement has been attributed to several causes.

EXTRANEOUS SUBSTANCES. 37

Vital Movements.—Place a drop of decomposing urine on a slide, cover and focus with H. The field contains innumerable minute spherules and rods (bacteria) which are in active motion, resembling somewhat the Brownian movement, although sufficiently distinctive after close observation.

After having rubbed the tongue for a moment against the inner surface of the cheek, put a drop of saliva on a slide, cover, and focus H. Among the numerous thin, nucleated scales and débris, small granular spherules—the salivary corpuscles—will be found. Select one of the last, centre, and focus H with extreme care. The minute granules within the cells are in active motion, resembling the Brownian movement; but with proper conditions the motion may continue for many hours.

EXTRANEOUS SUBSTANCES.

Before we begin the study of animal tissues, I wish to have you become somewhat familiar with the appearance of certain objects

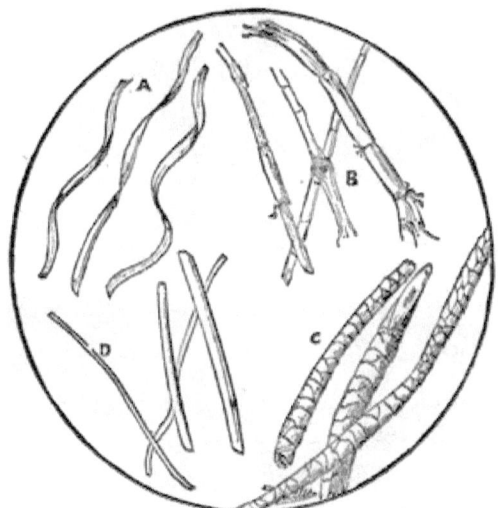

FIG. 21.—EXTRANEOUS SUBSTANCES.

A. Cotton fibres, showing the characteristic twist.
B. Linen fibres, with transverse markings indicating segments.
C. Wool. The irregular markings are produced by the overlapping of flattened cells. Wool may be distinguished from other hairs by the swellings which appear at irregular intervals in the course of the former.
D. Silk. Smooth and cylindrical.

which are frequently, through accident or carelessness, and often in spite of the utmost care, found mixed with our microscopical specimens. Among the more common objects floating in the air and gaining access to reagents, to subsequently appear in our mounted specimens, are the following:

Fibres.—Procure minute pieces of uncolored linen, cotton, wool, and silk. With a needle in either hand, tease out or separate a few fibres on slides, add a drop of water, and cover.*

Starch.—Procure samples of wheat, corn, potato and arrow-root starch, or scrape materials containing any one of these substances

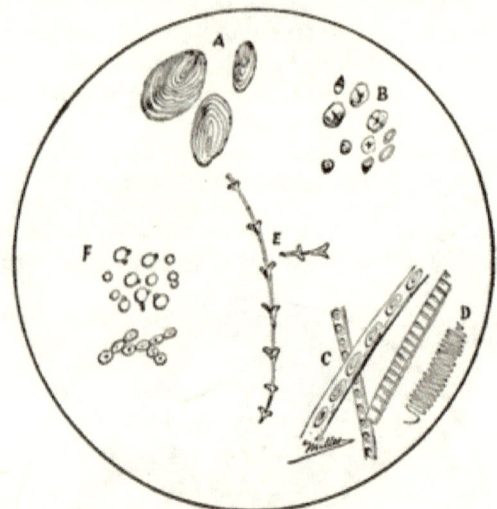

Fig. 22.—Extraneous Substances.

A. Granules of potato starch.
B. Corn starch.
C. Wood fibres. The circular dots are peculiar to the tissue of cone-bearing trees.
D. Spiral thread from a tea leaf.
E. Fragment of feather.

with a sharp knife. To a minute portion on the slide add a drop of water, cover, and examine L and H.

Wood Shavings, Feathers, Minute Insects, Portions of Larger

---

* These substances, as well as most of those which follow under the same heading, may be mounted permanently as follows: Put the dry material in clean turpentine for a day or two, to remove the contained air. Transfer to the slide, tease, separate, or arrange the elements, after which wipe away the turpentine with strips of blotting-paper. Add a drop of dammar and place the cover-glass thereon. The weight of the cover will be sufficient to press the object flat, if it be properly teased or separated. Although I do not advise the making of colored rings around cover-glasses, they may be formed after first protecting the dammer with a ring of gelatin (*vide* formulæ).

Insects, Pollen, etc., are easily mounted temporarily or permanently as above. They are very commonly found in urine after it has been exposed to the air, and their recognition is very important.

Let me urge you to become familiar with the microscopical appearance of the commoner objects which surround us in every-day life. The most serious mistakes have resulted from ignorance of this subject. Vegetable fibres have been mistaken for nerves (!) and urinary casts, starch granules for cells, vegetable spores for parasitic ova, etc.

STRUCTURAL ELEMENTS.

Certain anatomical structures, of a more or less elementary nature, are united in the composition of organs. These structural elements will with propriety first claim notice from us.

CELLS.

A typical cell is a microscopical sphere of protoplasm, constituted as follows (*vide* Fig. 23):

A. Limiting membrane.
B. Cell-body.
C. Nucleus.
D. Nucleolus.

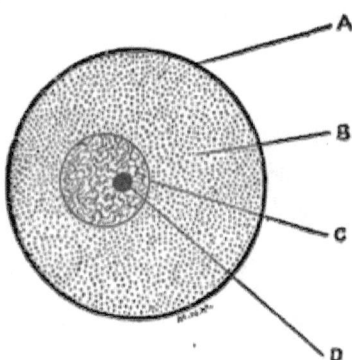

FIG. 23.—ELEMENTS OF A TYPICAL CELL.

The *wall* consists of an apparently structureless membrane of extreme tenuity.

The *cell-body* may be either clear (jelly-like), granular, or fibrillated.

The *nucleus* is a minute spherical vesicle, with a limiting membrane inclosing a clear gelatinous material, traversed by a reticulum of fibrillæ.

The *nucleolus* consists of a spherical granular enlargement upon the fibrillæ of the nucleus.

Deviations from the type are most frequent, and vary greatly as to form, number of elements, and chemical composition.

Fig. 24.—A Cell Nucleus, with Network and Nucleolus. Diagrammatic.

The typically perfect cell is rarely seen in human tissue on account of the length of time which commonly elapses between death and observation of the structure, the delicate fibrillæ of the nuclei usually appearing as a mass of granules.

CELL DISTRIBUTION.

The complex mechanism of the body had its origin in a single cell. This preliminary structure, endowed with the power of proliferation, became two cells. Two having been produced, they became four; the four, eight; and thus progression advanced until they became countless. Some of these cells remained as such; others altered in form and composition gave birth to muscle, bone, etc., etc. The study of these processes belongs to physiology.

The adult body is composed largely of cells of various forms. The different physiological processes, as secretion, absorption, respiration, etc., are effected through the intervention of these anatomical elements.

All free surfaces, within or without the body, *are covered with cells.* The entire skin, the outside of organs, as lung, liver, stomach, intestine, brain, etc., etc.; all cavities, as alimentary tract, heart, ventricles of the brain, blood-vessels, ducts, all present a superficial layer of cells.

VARIATION IN FORM OF CELLS.

Alteration from the typical or spherical form is effected mainly through pressure consequent upon active proliferation of contiguous cells, or growth of surrounding fibrous tissues.

FLAT CELLS.

If a cell be subjected to pressure on two opposite sides, a flattening ensues, and a scale-like element results. Flat cells are united to form a continuous structure in different ways.

SQUAMOUS, STRATIFIED, AND TRANSITIONAL EPITHELIUM.

The simplest method of tissue production by means of flat cells is that of superposition, constituting *squamous epithelium.* Cells are placed one over the other, generally without great regularity. If regular, and in several layers, the structure is called *stratified* epithelium; if only in a few layers, it is termed *transitional* epithelium. The superficial layer of the skin affords an example of

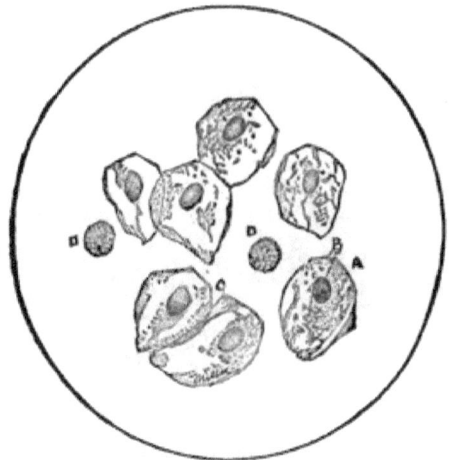

FIG. 25.—SQUAMOUS CELLS FROM BUCCAL EPITHELIUM
A. Typical cell. B. Its nucleus.
C. Union by overlapping forming laminæ.
D. Salivary corpuscles. × 400.

squamous, stratified epithelium. The bladder, pelvis of the kidney, and vagina are lined with transitional epithelium.

The thin, flat scales from the mouth may be demonstrated by scraping a drop of saliva from the tongue with the handle of a scalpel, transferring it to the slide, and applying the cover. The size of the drop of saliva should be carefully adjusted so as to fill the space between the cover-glass and slide. Too little will cause the cover to adhere so tightly to the slide as to press the cells out of form; too much, and the saliva flows over the cover and soils the objective. With a glass rod, place a drop of the dilute eosin solution on the slide, and with a needle lead it to the edge of the saliva. The dye will pass under the cover slowly; and, gradually, whatever anatomical elements there may be present will be stained. Observe that the nuclei of the flat scales first take the dye, and appear of a deep pink; while the other portions are either colorless or very lightly stained.

Find a typical field and sketch it with pencil, afterward tinting with dilute eosin.

PAVEMENT EPITHELIUM.

When thin, flat cells are disposed in a single layer, like tiles, the epithelium is termed *pavement* or tessellated. These cells are often quite regularly polygonal (although this obtains more frequently with tissue from the lower animals), and they are always connected by their edges by means of an albuminous cement.

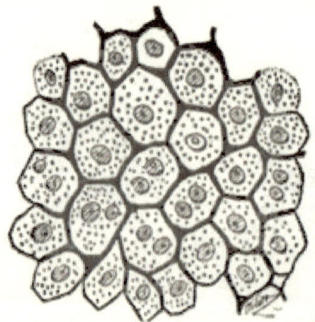

FIG. 26.—PAVEMENT EPITHELIUM. DIAGRAMMATIC.

This structure is very extensively distributed. Most serous surfaces—*e.g.*, the pleuræ, omenta, mesenteries, and peritoneal surfaces generally—are so covered. The lining of the heart, of arteries and veins, and of lymph channels is constructed with these cemented cells. Blood-capillaries are formed almost entirely of such elements.

The best demonstration is made by coloring the cement which unites the cells. If a tissue, covered with this epithelium, be placed for a few minutes in a solution of nitrate of silver a chemical union ensues; an albuminate of silver is formed which blackens in the light, thereby mapping out the cells with great precision and clearness.

It is nearly impossible to procure human tissue for this purpose, as the cement substance decomposes soon after death. The mesentery of the frog affords a good example of pavement cell structure; and differs but little from the arrangement on human serous surfaces.

Kill a large frog by decapitation, and open the abdomen freely by an incision along the median line. Pull out the intestines by grasping the stomach with the forceps. This will expose the small intestine, which you will remove, together with the attached mesentery, by means of quick snips of the scissors. Work as rapidly

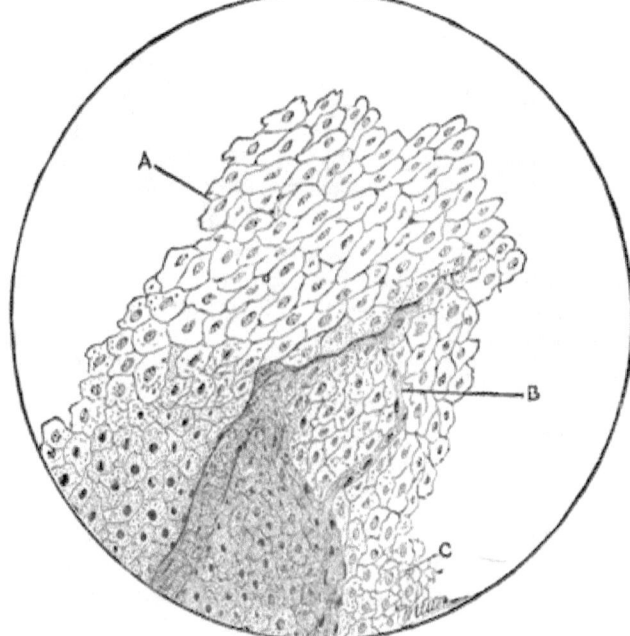

Fig. 27.—Pavement Epithelium from Frog's Mesentery. Silver Staining.

A. Area showing the outlining of the pavement cells by the silver-stained cement substance. The nuclei have been brought out by the carmine. Minute stomata may be seen between certain cells.

B. A blood-capillary terminating below in an arteriole. The silver has outlined the endothelia of the vessels.

C. An area showing both layers of the pavement. The deeper cells are faintly outlined, being out of focus. The silver has been deposited over the lower portion of the specimen, nearly obscuring the cement lines. × 250.

as possible and avoid soiling the tissue with blood. Throw the gut into a salt-cellar filled with silver solution (*vide* formulæ), where it must remain for ten minutes covered from the light. Lift the tissue from the solution by means of a strip of glass (or a platinum wire), and throw into a saucer of clean (preferably distilled) water, changing the latter repeatedly for ten minutes. After washing, and while yet in the water, expose to sunlight (perhaps fifteen minutes) until a brown tint is acquired which indicates the proper staining.

Proceed to stain the intestine, with mesentery attached, with borax-carmine as directed for sections, excepting that, as the mass is great, it must be washed twice in alcohol after the oxalic acid.

We have allowed the mesentery to remain connected with the gut, that the former might not curl, as it would have done had it been separate. The preparation having reached the oil of cloves, proceed with a pair of scissors to snip off a small, flat piece of mesentery. Remove it to a slide, clean off the oil, apply dammar, and cover.

The mesentery, you have learned from descriptive anatomy, constitutes a support for blood and lymph vessels which are in connection with the intestine. The vessels are held together with a little delicate fibrous (connective) tissue, and are covered above and below with a layer of pavement epithelium.

You will observe prominently some dark lines (the larger vessels) traversing the specimen. Select a thin spot between the vessels and focus H—you have a picture like Fig. 27. The field is traversed by very delicate dark lines, indicating the position of the cement substance; while the nuclei of the cells are pink from the carmine staining.

With the fine adjustment-screw run the tube of the microscope down carefully. The cement lines will disppear, and before they are completely out of focus, another set of cells will appear below the first set. So you may alternately bring into view the upper and under layers of cells covering the respective sides of the mesentery.

Observe the irregular shape of the cells. Note, also, that the cells on one side average larger than those on the other side. You may also notice in various parts of the specimen blood-vessels lined with cells which are outlined with the silver staining.

Sketch a field showing the elements as in Fig. 27, and stain the nuclei with the carmine solution.

COLUMNAR CELLS.

Columnar Epithelium.

Columnar cells are found, generally, throughout the alimentary and respiratory tracts. They also line the cerebral ventricles, the urinary and Fallopian tubes, the uterus, etc. This epithelium is quickly destroyed after death and is difficult of perfect demonstration except in an animal recently killed.

Procure from the abbatoir a portion of the small intestine and bronchus of a pig, and with the curved scissors snip out small pieces from the mucous surface of each. Macerate in one-sixth per cent of chromic acid for twenty-four hours.

Place a piece of the gut on a slide and, after having added a drop of the acid solution, scrape off the mucous surface with a knife and remove the remainder of the gut. Add a cover-glass and focus II. You will find cells in various conditions, from isolated examples to small groups like Fig. 28.

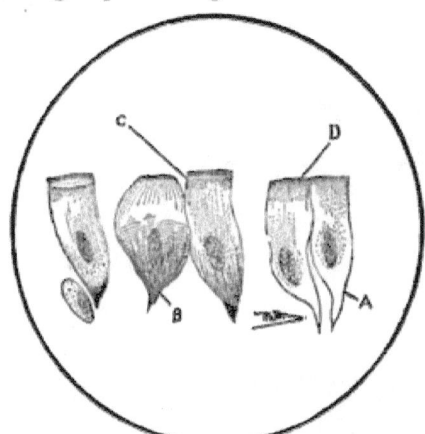

FIG. 28.—COLUMNAR CELLS FROM SMALL INTESTINE OF RABBIT.
A. Tapering attached extremity.
B. A swollen goblet cell.
C. Finely-striated free border.
D. Transparent line of union between the striated portion and the body of the cell.
× 400.

Observe that the attached ends of the cells are often small and pointed, and that spheroidal and ovoidal cells are frequently wedged in between them. Note the free border: it consists of striæ, and is separated from the body of the cell by a translucent line. This appearance is also that of the epithelium in the human intestine.

Ciliated Columnar Epithelium.

Prepare, by scraping, a slide from the mucous surface of the pig's bronchus (which has been macerating in the chromic acid).

Observe the cilia on the free border of the cells. Interspersed between ciliated cells, much-enlarged individuals may be found, the so-called beaker, goblet, or mucous cells.

The motion of the cilia may be demonstrated as follows:

Carefully open an oyster so as to preserve the fluid. On examination you will notice the leaflets, shown in Fig. 30, commonly called the beard. With the scissors snip off a fragment of the free border of this beard, add a drop of the liquid from the oyster, and tease with a pair of needles. Apply the cover and focus II.

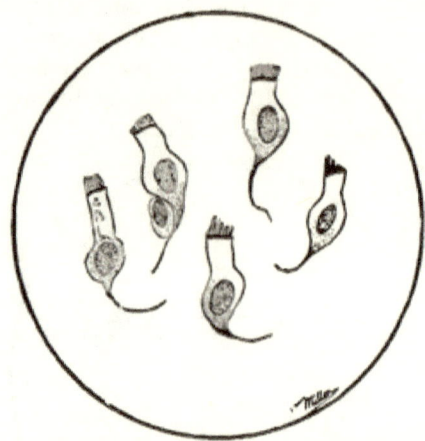

Fig. 29.—Ciliated Columnar Cells from Bronchus of Pig. × 400.

At first, the individual cilia cannot be demonstrated on account of their rapid vibration. After a few moments, however, the action becomes less energetic, and the hair-like appendages of the cells are to be plainly seen.

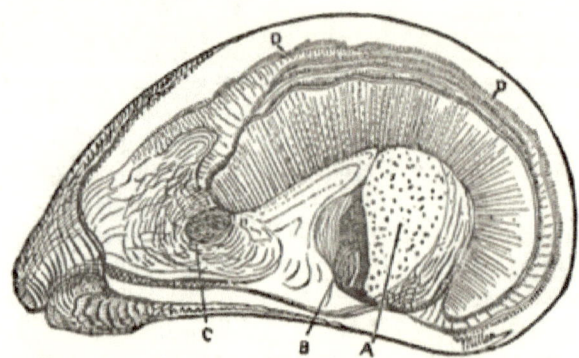

Fig. 30.—Oyster, opened to show Method of procuring Living Ciliated Cells.
A. The divided muscle. This must be sectioned before the shell can be opened.
B. The heart.
C. Liver.
D, D. The so-called "beard." These laminæ are covered with cells provided with cilia; and a fragment of the free border of one of the leaflets may be snipped with the scissors and examined as described in the text.

Of course none of the above objects are intended to be permanent.

RED BLOOD-CORPUSCLES. 47

SPHEROIDAL CELLS.

The only cells which have, in any very great number, retained their primitive spheroidal form are the corpuscles of the blood and of the lymphatic system.

In solid organs, the cells, primarily spheroids, often become polyhedral from pressure.

Cells, developed spheres, not infrequently send out prolongations, forming either *stellate* or *polar* cells according to the size of the radiating processes.

RED BLOOD-CORPUSCLES.

The human red blood-corpuscle is a flattened, bi concave disc, one-three-thousand-two-hundredth of an inch in diameter. It

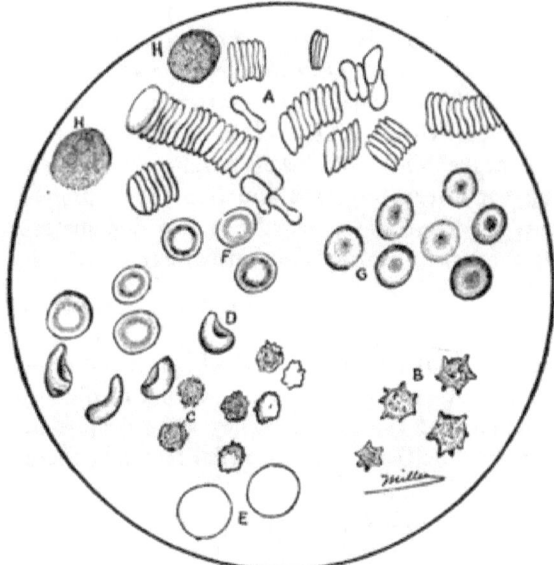

FIG. 31.—CORPUSCULAR ELEMENTS OF HUMAN BLOOD.
A. Colored corpuscles adhering by their sides—*rouleaux.*
B. The same crenated.
C. The same shrunken.
D. The same having absorbed water.
E. The same still more swollen.
F. The same with the plane C D, Fig. 32, in focus.
G. The same with the plane A B, Fig. 32, in focus.
H. Colorless corpuscles.   × 400.

presents a mass of protoplasm destitute, as far as the microscope shows, of nuclei, cell-wall, or any structure whatsoever.

Certain changes in form result, after removal from the circulation, viz.: 1. They may adhere by their broad surfaces forming columns. 2. From shrinkage they may become crenated. 3. Still further shrinkage produces the chestnut-burr appearance. 4. From absorption of water they may swell irregularly, obliterating the concavity of one side. 5. From continuous absorption they swell, forming spheres which are finally dissolved.

Wind a twisted handkerchief tightly around the left ring-finger;

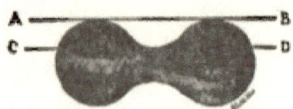

FIG. 32.—DIAGRAM OF A COLORED BLOOD-CORPUSCLE; SIDE VIEW, SHOWING THE BI-CONCAVITY.
A, B. Upper plane; which, in focus, gives the appearance shown at G, Fig. 31.
C, D. Plane giving the appearance F, Fig. 31.

prick the end with a clean needle, and squeeze a minute drop of blood on a slide, add a drop of saliva, cover and focus H.

Observe: 1. That considerable variation in size of the red blood-corpuscles exists. 2. The color—a delicate straw tint. 3. That the concave centre of the corpuscles which lie flat can be made to appear alternately dark and light according to the focal adjustment. 4. That the concavity is also demonstrated as the corpuscles are turned over by the thermal currents.*

BLOOD-PLATES.

Minute corpuscular elements in the blood about one-fourth the size of the red discs exist in the proportion of about one of the former to twenty of the latter. They are colorless ovoid discs; and are regarded by Osler as an essential factor in the coagulation of the blood.

Prick the thoroughly clean finger with a needle. Over the puncture place a drop of solution of osmic acid (*vide* formulæ) and squeeze out a minute drop of blood, so that, as it flows, it is covered by the acid solution. This fixes the anatomical elements, providing against further change. The blood, as soon as drawn, must, with the acid, be immediately transferred to a slide and cov-

---

* The student is at this time advised to study the corpuscular elements of the blood of such animals as he may be able to command. The red corpuscles of mammals (excepting the camelidæ) do not vary in appearance from those of man, excepting in size. Those of birds, fishes, and reptiles are elliptical with oval nuclei. Corpuscles of the blood of invertebrates are not colored.

ered. To provide against evaporation, run a drop of sweet oil around the edge of the cover.

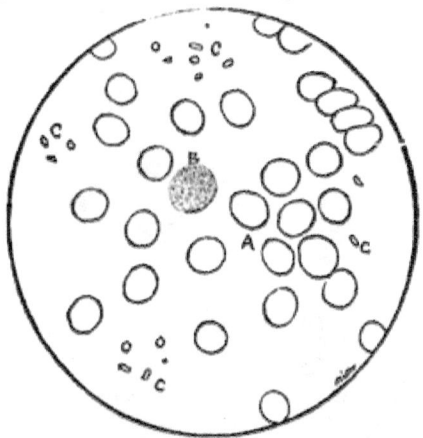

Fig. 33.—HUMAN BLOOD PRESERVED WITH OSMIC ACID.
A. Colored corpuscles.   B. Colorless corpuscle.
C, C, C. Groups of plaques.   × 400 and reduced.

The blood-plates may be found, after careful search, bearing the relation to the red corpuscles seen in Fig. 33.

WHITE OR COLORLESS BLOOD-CORPUSCLES.

The white blood-corpuscle is a typical cell, spherical in form, presenting generally a nucleus—often two or more—with nucleoli. In diameter about the one twenty-five-hundredth of an inch, they are usually found in the blood in proportion of one to three hundred to one thousand red corpuscles. The nucleus of the white corpuscle possesses nearly the same refractive index as the body of the cell, and is therefore difficult of demonstration without the use of reagents or staining.

Procure a drop of pus from a healing wound, mix it on a slide with an equal quantity of dilute eosin solution, cover, and examine it.

Pus is colorless, containing spherical nucleated corpuscles, the perfect ones resembling exactly those found in healthy blood. Observe that the nuclei, some cells containing three or even four, are stained with the eosin. Minute pigment-granules and fat-globules appear in many of the pus-cells, and others are broken and distorted.

POLYHEDRAL CELLS.

With a scalpel scrape the cut surface of a piece of liver from a recently-killed pig; place a minute portion of the finer part on a slide; add a drop of normal salt solution (*vide* formulæ); mix with a needle, and put on the cover-glass.

With II observe, among the numerous blood-corpuscles, fat-globules, etc., the *polyhedral liver cells*, about twice or three times

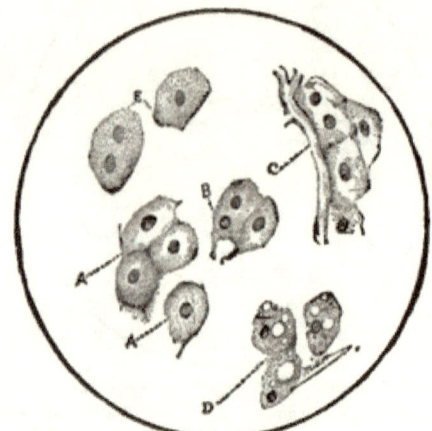

Fig. 34.—GLANDULAR EPITHELIA.

A, A. Polyhedral cells from human liver.
B. Double nuclei.
C. Cells from same showing connection with a capillary.
D. Same cells infiltrated with globules of fat.
E. Cells from liver of pig showing intracellular network. × 400.

the diameter of a white blood-corpuscle (Fig. 34). Notice the large spherical nuclei, with nucleoli. Note, also, the yellow pigment-granules and the fat-globules in the body of the cells. Masses of these cells resemble somewhat pavement epithelium; they are not flat but polyhedral.

STELLATE CELLS.

When we arrive at the study of the skin, I shall show you some very beautiful examples of stellate cells. I prefer to leave their demonstration until you have become more familiar with tissue-cutting.

POLAR CELLS.

As I have stated, spheroids may send off processes. These prolongations may be one, two, three, or more in number, constituting *unipolar*, *bipolar*, *tripolar*, etc., cells. The best demonstration is made from the nervous system, where these poles are continued as nerves, etc.

WHITE FIBROUS TISSUE. 51

From a freshly-slaughtered ox, sheep, or pig (the first being the best) obtain a piece of the spinal marrow from the region of the neck. Cut it transversely into discs about one-eighth of an inch thick, and place them in the chromic-acid fluid diluted with an equal bulk of water. After thirty-six hours, place one of the pieces in water, and with a needle pick out minute fragments from the anterior horn of the gray matter (refer to the diagram of the spinal cord) and transfer them to a slide. Add a drop of water and break the tissue into minute fragments by teasing with a pair of needles. Examine from time to time with L, to note the progress of the teasing. When properly teased, put on the cover-glass and search for large nucleated cells from which the prolongations or horns are given off. Compare with Fig. 123.

Cells may be classified as follows:

Epithelial—covering-cells, as in skin.
Endothelial—lining-cells, lining vessels or cavities.
Glandular—constituting the parenchyma of organs.

CONNECTIVE (FIBROUS) TISSUES.

Certain elementary structures of similar origin and mode of development, and serving alike to unite the various parts of the body, have been termed connective tissues. Custom has restricted the term, in its every-day employment, so as to apply to white fibrous tissue or, at least, to tissue which always resembles this more closely than any other, and I shall so use the expression in this work.

WHITE FIBROUS TISSUE.

This, the connective tissue *par excellence*, is composed of exceedingly fine fibrillæ (one-fifty-thousandth of an inch), which are aggregated in irregularly-sized and variously-disposed bundles. It forms long and exceedingly strong tendons connecting muscle and bone; its fibres interlace, forming the delicate network of areolar tissue; it forms thin sheets of protecting and connecting aponeuroses; or, supporting vessels, it permeates organs and sustains the parenchyma of glands.

The fibres are held together by means of a transparent cement, which may be softened or dissolved in acetic acid. They may exist, as in dense tendons, without admixture.

Cells are found between the bundles of fibres, known as connective-tissue corpuscles or fibro-blasts. The older and more dense the structure, the less frequent are these cells; while in young (recent) connective tissue, stained, the nuclei of the corpuscles constitute a prominent feature of the specimen under the microscope.

Having obtained a piece of tendon from a recently-killed bullock, tease a fragment on a slide in a few drops of water. Select a portion which splits easily and separate the fibrils as much as possible. Cover and examine it.

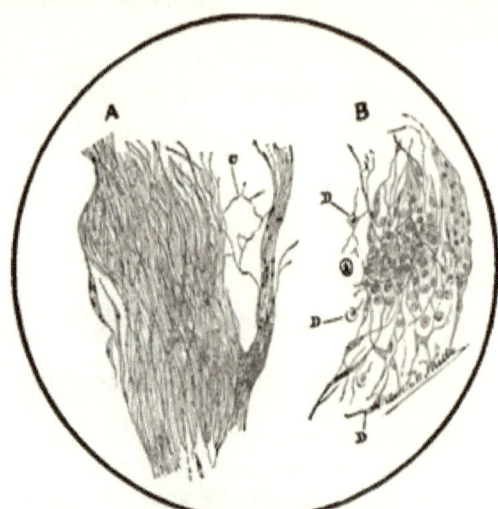

Fig. 35.—Connective Tissue.

A. Teased fibres from a tendon.
B. New connective tissue from a cirrhotic liver.
C. Fibrillæ.
D. Elongate cells in last, showing mode of formation of fibrillæ from cell elements. × 400.

Fine, wavy fibres are seen composing the fasciculæ. If the dissection has been sufficiently minute, you may succeed in demonstrating ultimate fibrillæ. These are best made out, as at C in Fig. 35, where the parts of a bundle have been separated for some distance, leaving the finer elements stretching across the interval.

B in Fig. 35 shows recently-formed connective tissue from the liver, where this structure had so increased as to largely obliterate the parenchyma of the organ.

YELLOW ELASTIC TISSUE.

This tissue consists of coarse shining fibres (averaging one-three-thousandth of an inch) which frequently branch and anastomose. They are highly elastic. Under the microscope the fibres are colorless; but when aggregated, as in a ligament, the mass is yellow.

YELLOW ELASTIC TISSUE. 53

FIG. 36.—TEASED YELLOW ELASTIC TISSUE FROM THE LIGAMENTUM NUCHÆ. × 250.

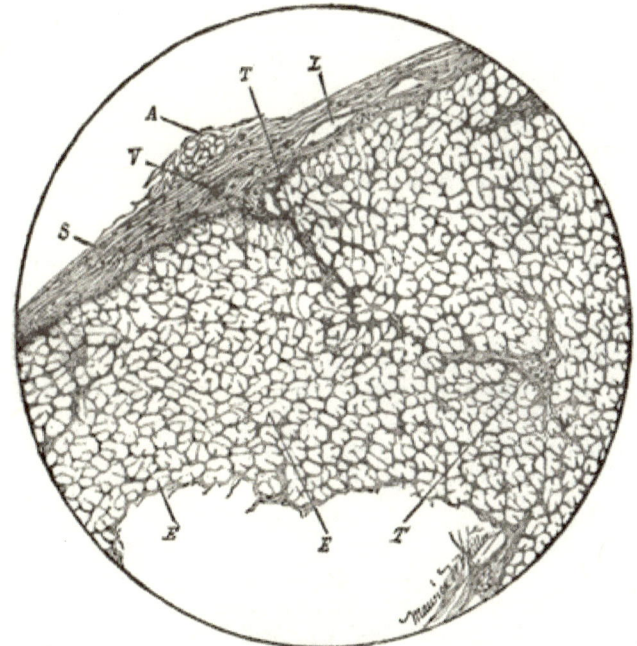

FIG. 37.—TRANSVERSE SECTION OF PART OF THE LIGAMENTUM NUCHÆ.

S. Sheath of the ligament, sending prolongations within—as at T, T—dividing the structure into irregular bundles or fasciculæ.
   L. Lymph spaces in the connective tissue.
   A. Adipose tissue in the sheath.
   V. Blood-vessels in transverse section.
   E, E. Primitive fasciculæ of yellow elastic tissue fibres. × 250.

Procure a small piece of the *ligamentum nuchæ* of the ox, and tease it on the slide after its having been macerated in acetic acid for a few moments. The acid softens the fibrous connective tissue and facilitates the teasing process.

The individual fibres having been isolated, they appear as in Fig. 36. When broken, they curl upon themselves like threads of India-rubber.

This tissue is variously disposed throughout the body where great strength with elasticity becomes necessary. The large arteries are abundantly supplied with elastic fibre, arranged in plates, in alternation with muscle. As a network, it is mixed with connective tissue in the skin, and in membranes generally. It contributes elasticity to cartilage where the fibres form an intricate network.

Ligaments are composed largely of yellow elastic tissue. Fig. 37 is drawn from a portion of a stained transverse section of the *ligamentum subflava*.

A strong *sheath* of fibrous tissue is thrown around the whole ligament, a portion of which is seen at S. This sheath sends prolongations, T, T, into the structure, dividing it into irregular bundles, which support nutrient vessels. The elastic fibres seen in transverse section, as at E, E, are observed strongly bound together with fibrous tissue, which penetrates the smaller fasciculæ, dividing them into the *ultimate fibrillæ*.

ADIPOSE TISSUE.

Adipose or fat tissue is a modification of and development from ordinary connective tissue.

It originates in certain contiguous connective-tissue corpuscles, becoming filled with minute fat-globules. These ultimately coalesce and form single, large globules, which bulge out the cell-bodies until they become spheroids; the nuclei at the same time are displaced to the periphery. An aggregation of such cells forms a lobule of adipose tissue. The cells are often so closely packed as to assume a polyhedral form. From malnutrition, this fat may be absorbed, ordinary connective tissue remaining.

You will bear in mind the fact that whenever fat exists in a condition of minute subdivision, the particles always assume the globular form; and that while adipose tissue contains fat, fat alone is not adipose tissue.

ADIPOSE TISSUE.

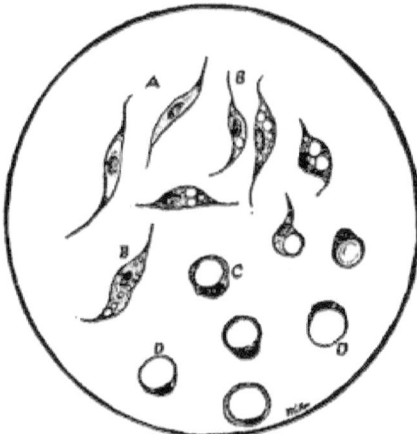

Fig. 38.—Connective-tissue Cells Containing Fat—indicating the Mode of Formation of Adipose Tissue.
  A. Ordinary elongate connective-tissue cells.
  B. Same containing minute globules of fat.
  C. Coalescence of the fat-globules and displacement of the nucleus.
  D. Still greater increase of the fat. × 400.

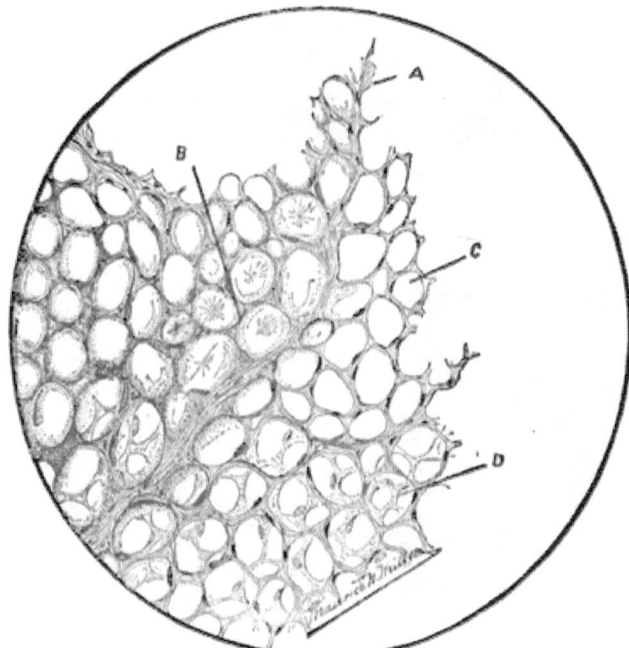

Fig. 39.—Adipose Tissue from Teased Human Omentum. Stained with Hæma.
  A. Connective-tissue framework.
  B. Cells distended with fat, showing fat-crystals.
  C. Cells from which the fat has been dissolved by ether.
  D. Cells faintly seen below the more sharply focussed plane. × 400.

CARTILAGE.

Cartilage consists of a dense basis substance, in which cells or chondroblasts are imbedded. It presents in three forms.

HYALINE CARTILAGE.

The matrix of hyaline cartilage is translucent, dense, and apparently structureless. Minute channels in certain instances and delicate fibrillæ in others have been demonstrated.

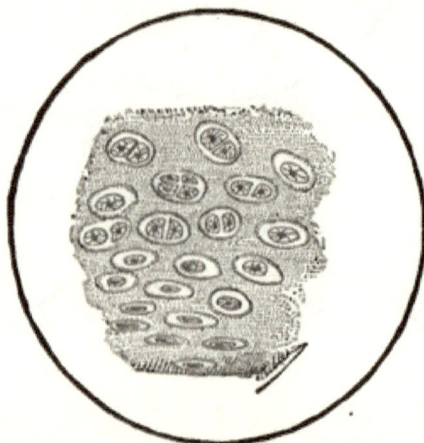

FIG. 40.—SECTION OF HYALINE CARTILAGE FROM A HUMAN BRONCHUS.
The ground-substance is apparently structureless, and it contains the membrane-lined excavations in which one, two, three, or more cartilage cells appear. These cells show a well-marked intracellular network. ×400.

The basis material contains excavations, generally spherical, called *lacunæ*. They are lined with a delicate membrane and contain one, two, three, and perhaps as many as eight cells—the *cartilage corpuscles*.

Hyaline cartilage is found covering joints generally, where it is termed articular cartilage. It is also found in the trachea, the bronchi, the septum narium, etc.

Fig. 40 shows a section from one of the rings of a large bronchus.

FIBRO-CARTILAGE.

Fibrous connective tissue predominating largely in the basis substance produces a structure of great strength—fibro-cartilage

The intervertebral discs afford an example of this variety, from a section of which Fig. 41 has been drawn. The membrane lining

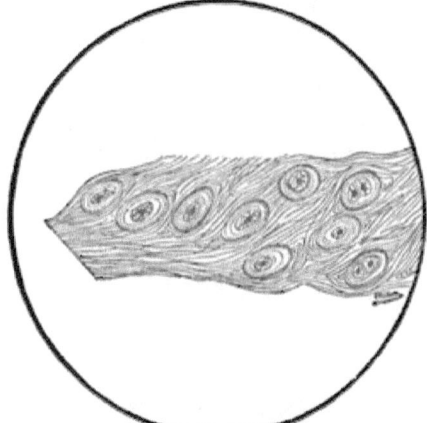

Fig. 41.—Fibro-cartilage from an Intervertebral Plate or Disc.
The ground-substance, unlike that of the hyaline variety, consists of dense fibrous tissue with little calcareous matter. × 400.

the lacunæ is much thicker than in the previous example, and the fibrous tissue is a very prominent feature of the ground-substance.

ELASTIC OR RETICULAR CARTILAGE.

As the name implies, yellow elastic tissue is an important element of the ground substance of elastic cartilage. It presents in

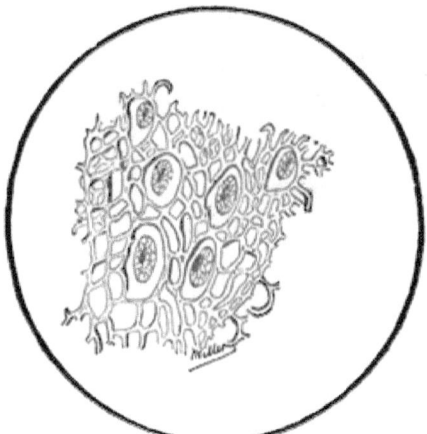

Fig. 42.—Elastic Cartilage from Ear of Bullock.
The ground-substance consists largely of a network of coarse, yellow elastic tissue. × 400.

the form of a reticulum, as shown in Fig. 42. It is not extensively distributed in the human being, although the cartilages of the external ear, Eustachian tube, etc., are of this variety.

Cartilage should be hardened by the chromic acid and alcohol process. The sections from which the illustrations have been drawn were cut without the microtome. They should be cut extremely thin, not necessarily large. We frequently succeed in getting good fields from the *thin edges* of sections which may be elsewhere too thick. Stain with hæma. and eosin. The differentiation will be excellent. The delicate nutritive channels in the matrix connecting the lacunæ may be demonstrated in the cartilage of the sternum of the newt; the xiphoid appendix is sufficiently thin without sectioning.

BONE.

Bone consists of an osseous, lamellated matrix, in which occur irregularly-shaped cavities—*lacunæ*. The latter are connected by

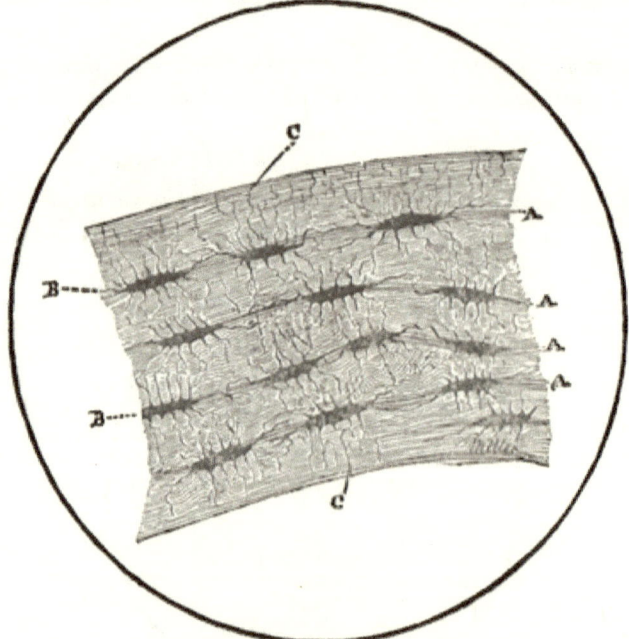

FIG. 43.—PORTION OF A TRANSVERSE SECTION FROM A DRIED FEMUR, SHOWING PART OF THE WALL OF AN HAVERSIAN SYSTEM.

A, A. Bony laminæ.
B, B. Lacunæ.
C, C. Canaliculi. × 400.

means of exceedingly fine channels—*canaliculi*. The lacunæ contain the *bone corpuscles*, the bodies of which are projected into the canaliculi.

In compact bone, the blood-vessels run in a line parallel with the long axis of the bone, in branching inosculating channels (averaging one-five-hundredth of an inch)—the *Haversian canals*. The lamellæ of osseous tissue are arranged concentrically around these canals. A single Haversian canal with the lamellæ surrounding and belonging to it constitute an *Haversian system*.

Fig. 44.—Transverse Section of Portion of a Dried Long Bone, showing the Haversian Systems.

A, A, A. An Haversian system.
B. Haversian canal.

The lacunæ, canaliculi, and Haversian canals all appear black in the section, as they are filled with air and the bony fragments resulting from grinding of the section. × 60.

The lamellæ beneath the periosteum are not arranged as above, but parallel with the surface of the bone. These plates are perforated at right angles, and obliquely by blood-vessels from the periosteum, as they pass on their way to the Haversian canals. These lamellæ are also perforated by calcific connective tissue—the *perforating fibres of Sharpey*.

An Haversian canal contains (Fig. 44) an artery, a venule, lymph channels, and a nerve filament. The whole is supported by connective-tissue cells with delicate processes. The walls of the lymph spaces are prolonged into the canaliculi and thus placed in connection with the elements of the surrounding lacunæ.

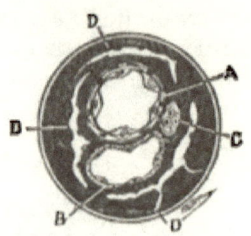

FIG. 45.—DIAGRAM OF AN HAVERSIAN CANAL.
    A. Artery.
    B. Vein.
    C. Nerve.
    D, D, D. Lymph channels.

Each lacuna contains a bone corpuscle, the protoplasmic body of which sends prolongations into the contiguous canaliculi. In the adult bone, the cell is shrunken; and the processes just mentioned are not readily demonstrable.

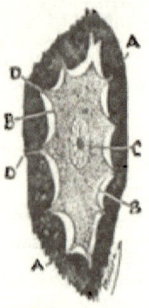

FIG. 46.—DIAGRAM OF A BONE LACUNA.
    A, A. Ground-substance of the bone.
    B, B. Limiting membrane of the bone corpuscle within the lacuna.
    C. Nucleus and nucleolus of the corpuscle.
    D, D. Projection of the cell-body into the canaliculi.

Fig. 44 has been drawn from a section of dry bone which has been sawn as thin as possible, and afterward rubbed down on a hone with water. It is a tedious process, and shows little but the osseous matrix. Bone should be decalcified for microscopical work, and it may then be readily cut in thin sections with a razor. The process is as follows:

To four ounces of the dilute chromic-acid solution add a drachm of C. P. nitric acid. The bone, previously divided into slices not over one-fourth of an inch in thickness, is then placed in the fluid, and should be completely decalcified in a week or ten days. Examine the pieces after twenty-four hours by puncturing with a needle. Should the action proceed too slowly, add a few drops more of the nitric acid from time to time. The bone eventually takes on a green color. After complete decalcification, wash the pieces for twenty-four hours in clean water, and preserve them, until required, in "B" alcohol. Small pieces of young bone may be decalcified in a saturated aqueous solution of picric acid. The process is slow, but it leaves the tissue in excellent condition.

Sections cut in the usual way may be stained with carmine and picric acid, and examined in a drop of glycerin. They should not after the staining be placed in the oil of cloves, as they would curl and become hard. Transfer them to equal parts of glycerin and water, from which they are to be carried to the slide. Add a drop more of the dilute glycerin if necessary and put on the cover-glass, carefully avoiding air-bubbles. If you desire to make a permanent mounting, the edge of the cover must be cemented to the slide.

Thoroughly wipe the slide, around the cover, with moistened paper, *until every trace of glycerin is removed*. Then with a sable brush, paint a ring of zinc cement (*vide* formulæ) around the slide just touching the edge of the cover-glass. Repeat the cementing in twenty-four hours. A turn-table will be a useful aid in this work. Dr. Carl Heitzmann, who uses glycerin as a universal mounting fluid, prefers ordinary black (asphalt) varnish as a cement.

SPECIAL CONNECTIVE TISSUES.

Connective Tissue of the Lymphatic System.—The matrix of lymphoid or adenoid tissue consists of a network of branching cells, which support the lymph-corpuscles. It is distributed extensively in organs, and where it appears in stained sections, the lymphoid cells are so numerous as to obscure the reticulum almost entirely. The structure will be minutely described in connection with the lymphatic system.

The Connective Tissue of the Central Nervous System (neuroglia) consists of branched connective-tissue cells, which are supported in an intimate network of exceedingly fine elastic fibrillæ, and will receive attention later in our work.

Embryonic Connective Tissue presents a homogeneous, mucoid matrix containing branched cells. It is not found normally in the adult.

MUSCULAR TISSUE.

This tissue is found in three varieties: 1. Non-striated or involuntary; 2. Striated, red, skeletal, or voluntary; 3. Cardiac.

NON-STRIATED MUSCLE.

The histological element of non-striated muscle is a spindle-shaped cell from one-tenth to one-five-hundredth of an inch long. The cell body presents longitudinal striæ, and contains an ovoid nucleus. The nucleus contains a reticulum which is probably in

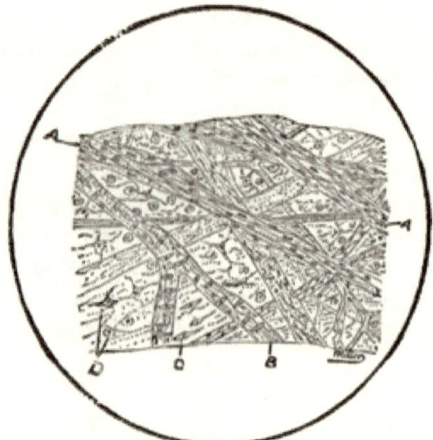

Fig. 47.—Wall of the Frog's Bladder, Stained with Hæma.

A, A. Bands of involuntary muscular fibre, recognized by the spindle-cell sarcous elements.
B. A small arteriole, showing the same muscular element.
C. Scattering muscle cells.
D. Connective-tissue cells. × 400.

connection with the fibrillæ, which produce the longitudinal striation of the body. The cells are not infrequently bifid at one or both extremities. A transparent cement substance serves to unite these cells in forming, with connective tissue, broad membranous plates, bundles, plexuses, etc. It serves to afford contractility, especially to the organs of vegetative life.

Kill a good-sized frog by decapitation, and open the abdomen on the median line. Fill the bladder with air, after the introduction of a blow-pipe into the vent. Remove the inflated bladder with a single cut with the curved scissors, and place it in a saucer of water. Proceed to brush it, under the water, with two camel's-hair pencils so as to remove all of the cells from the inner surface. It will bear vigorous rubbing with one of the brushes, holding it at the same time with the other. Transfer to alcohol for ten minutes, and afterward stain with hæma. and eosin. While in the oil, cut the tissue into small pieces, and mount flat in dammar. Examine L. and H.

Observe the bands of involuntary muscle crossing in various directions. You will distinguish (Fig. 47) between the muscle and the connective-tissue cells by their nuclei.

STRIATED MUSCULAR TISSUE.

A skeletal or striated muscle consists of cylindrical fibres, one-three-hundredth to one-six-hundredth of an inch in diameter, and one to two inches long. These primitive fibres are supported by a

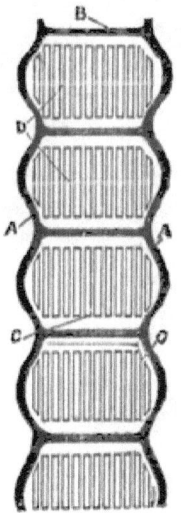

FIG. 48.—DIAGRAM INDICATING THE MINUTE STRUCTURE OF STRIATED MUSCULAR FIBRE.
A, A. Sarcolemma.
B. Krause's line connecting with the sarcolemma and dividing the fibril into compartments.
C, C. The rod-like contractile substance.
D. Hensen's middle disc.

delicate, transparent sheath—the *sarcolemma*. They are aggregated, forming *primitive fasciculi*, which are again united to form the larger bundles of a complete muscle. The connective tissue uniting the primitive fibres is termed *endomysium;* while that uniting the primitive bundles is the *perimysium*.

The primitive muscular fibres exhibit marked cross striations

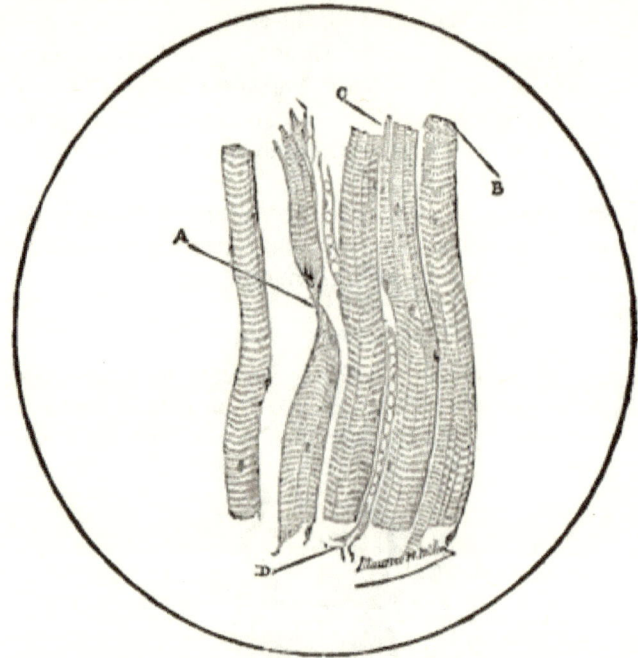

FIG. 49.—STRIATED MUSCULAR FIBRES FROM THE TONGUE, TEASED AND STAINED WITH HÆMA.
  A. A fibre, with the muscle substance wanting, from stretching during the teasing, the sarcolemma alone remaining.
  B. Partly separated disc of Bowman.
  C. Ultimate fibrillæ.
  D. A blood-capillary. × 400.

with faint longitudinal markings, the former being produced by alternate dark and light spaces.

Fig. 48 illustrates diagrammatically the theory of the structure of a primitive fibre: A indicates the *sarcolemma*. The dark substance B, B (*Krause's membrane*) divides the fibre completely, and is united with the sarcolemma. The light spaces C, C, between Krause's membranes, containing the *contractile substance*, are termed the *muscular compartments* or discs of Bowman. This contractile substance in the living muscle is semi-fluid, but in hardened tissue it can be split up, as indicated at C, into *rods*, the *sarcous elements*. A transparent line, D, in this contractile substance can sometimes be demonstrated; it is known as *Hensen's middle disc*.

Macerate human muscle, preferably that from the tongue, in dilute chromic acid for twenty-four hours; wash, tease in water, cover, and focus II. Fig. 49 was drawn from such a preparation.

The sarcolemma is best seen where the contractile substance has been broken. The muscle nuclei are seen at various points beneath the sarcolemma. Portions of a fibre have been split off transversely in places, indicating the *discs of Bowman*. Sarcous elements are indicated where the fibre has been split during the teasing. The capillaries are arranged in a direction parallel to the fibres, with frequent transverse connections.

CARDIAC MUSCULAR FIBRE.

It presents the following characteristics:
1. The fibres are smaller than those of ordinary skeletal muscle.
2. They are striated both transversely and longitudinally.
3. They branch, forming frequent inosculations.

Fig. 50.—Teased Cardiac Muscular Fibre.
Stained with hæma. × 400 and reduced.

4. They are divided by distinct transverse lines into short prisms.
5. Their nuclei are situated within the fibre.
6. They present no distinct sarcolemma.

Fig. 50 has been drawn from fresh cardiac muscle, teased in normal salt solution and tinted with eosin.

BLOOD-VESSELS.

Blood-vessels include *arteries, arterioles, capillaries, venules,* and *veins.* They are all lined with flattened endothelial cells cemented by their edges; and their walls are constructed from non-striated muscular, yellow elastic and fibrous connective tissues, in proportions varying according to the size and function of the vessel. Arteries are the active, while the veins are comparatively passive agents in the circulation of the blood.

The large arteries are eminently elastic, from preponderance of

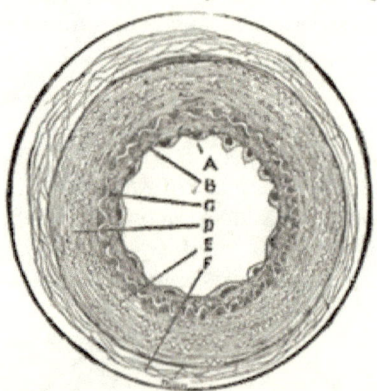

FIG. 51.—TRANSVERSE SECTION OF A MEDIUM-SIZED ARTERY. PARTLY DIAGRAMMATIC.
 A. The endothelial cells in profile.
 B. Elastic and connective tissue supporting the endothelium.
 C. The internal elastic lamina or fenestrated membrane. A, B, and C constitute the INTIMA of the artery.
 D. The MEDIA. It consists of muscular and elastic tissues in alternating layers.
 E. Points to one of the elastic layers.
 F. The ADVENTITIA. Loose connective tissue, with few elastic fibres.

yellow elastic tissue; while the arterioles are eminently contractile, from excess of muscular fibre.

Arteries possess three coats: the *intima* (internal), *media* (middle), and *adventitia* (external).

Fig. 51 represents a medium-sized typical artery. The intima, or internal coat (1), consists of a layer of flattened endothelial cells, which rest upon fibrous connective tissue, with a few elastic fibres. The last is surrounded by a layer of elastic tissue, the elastic lamina or *fenestrated membrane,* which is the external limit of the intima. It presents in a transverse section as a wavy (from contraction of the media) shining line; and is an important element, from its relation to certain abnormalities of the blood-vessels. The media (2) consists of alternate layers of elastic and muscular tissue. The adventitia (3) is composed of fibrous connective tissue, containing some elastic elements.

As we approach the larger arteries, the muscular tissue diminishes in quantity and the elastic tissue is increased. On the other hand, the elastic element diminishes with preponderance of muscle as we approach the smaller arteries, until we meet the arterioles, the walls of which are made almost exclusively of involuntary muscular fibre.

The walls of capillaries consist of a single layer of flattened endothelial cells cemented by their edges. The union is not quite

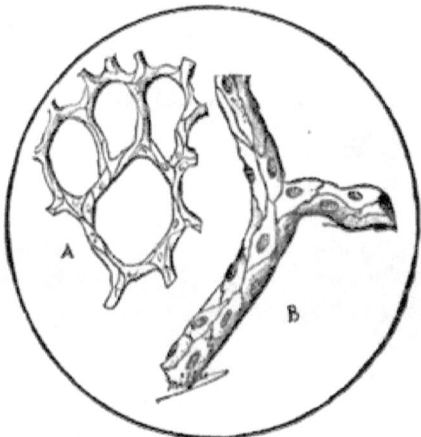

Fig. 52.—Isolated Blood-capillaries.

A. Plexus from a pulmonary alveolus, stained with silver. × 350.
B. Capillary from omentum, stained with silver and hæma. × 700.
In A the cells are outlined by the silver; while in B the nuclei in addition are brought out by the hæma.

continuous, as minute openings (stomata) are to be seen at irregular intervals.

The walls of veins are much thinner than those of arteries. The intima presents an endothelial lining, but no fenestrated membrane; and the line of demarcation between this coat and the media is often indistinct. The media contains muscular, but little elastic tissue; and the adventitia, usually the most prominent of the three coats, is composed largely of fibrous connective tissue.

I shall defer the microscopical examination of blood-vessels until we meet them in future sections of organs, as they are best studied in such connection.

PART THIRD.

ORGANS.

THE SKIN.

The skin consists of (1) the *epidermis* (or scarf skin), which everywhere covers and protects (2) the *derma* (corium or true skin).

The *epidermis* varies greatly in thickness in different locations; and in the thicker portions several layers may be differentiated. It is composed entirely of cells, while the derma is fibrous.

Epidermal Layers.
{
1. *Stratum Corneum,*
2. *Stratum Lucidum,* } Horny Layer.
3. *Stratum Granulosum,*
4. *Stratum of Prickle Cells,* } Malpighian Layer or
5. *Stratum of Elongate (Pigment)* } Rete Mucosum.
  Cells,
} Epidermis.

The *stratum corneum* consists of old, exhausted, flattened, and desiccated cells, which are constantly falling from the entire surface of the body. Dandruff consists of impacted cells from this source. Those portions most frequently exposed to friction—*e.g.*, the palms of the hands and soles of the feet—are protected by a corneous epidermal layer of great thickness.

The *sratum lucidum*, or clear layer, presents cells in form not unlike those in the preceding stratum; they are, however, translucent. This is properly a part of the previous stratum, is often absent, and frequently very difficult of demonstration.

The *stratum granulosum*, or granular layer, is composed of flattened cells containing opaque granules.

Immediately beneath the last-named layer, the cells become

strikingly altered in form and appearance. The *prickle cells* are polygons or compressed spheroids, with large, oval nuclei, and minute, projecting spines. By means of these processes they are very firmly united.

The fifth and last (deepest) layer of the epidermis is composed of a single rank of elongate cells, placed with their long axes at right angles to the surface of the skin. These cells contain the

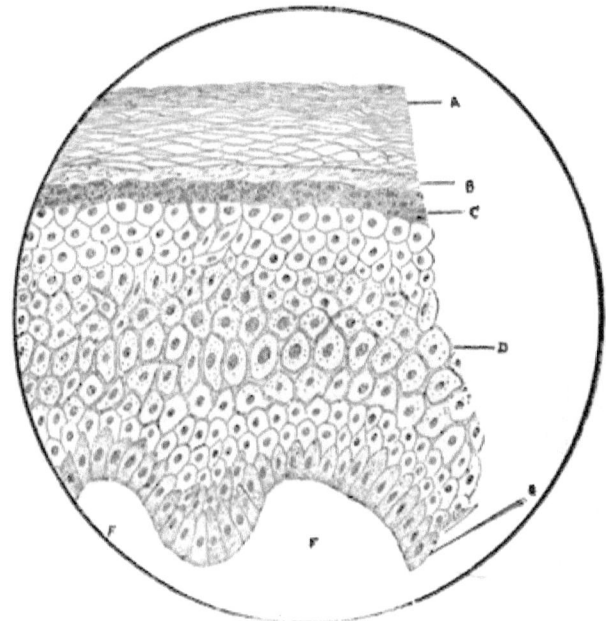

Fig. 53.—Vertical Section of the Epidermis from the Palm of the Hand. Stained with Hæma. and Eosin.

- A. Stratum corneum.
- B. Stratum lucidum.
- C. Stratum granulosum.
- D. Prickle cells of rete mucosum or R. Malpighii.
- E. Stratum of elongate cells, the lower limit of the epidermis.
- F, F. Indicate the position of two papillæ of the true skin or derma. × 400.

pigment which gives the hue peculiar to the skin of colored individuals.

The first two layers of the epidermis constitute, properly, the horny layer; while the remaining three strata compose the *rete mucosum* or *rete Malpighii*.

The *derma*, corium or true skin, is composed of dense, fibrillated connective tissue, so formed as to present minute elevations or papillæ over the entire surface of the body. These papillæ are covered with a basement membrane, and are protected from undue irritation by the epidermal layers.

The *subcutaneous cellular tissue* (upon which the true skin rests) consists of fibrillated connective tissue with elastic elements, from which strong interlacing bands are formed. These, in the deeper parts, form septa which support lobules of adipose tissue. These isolated collections of adipose tissue, when elongated and

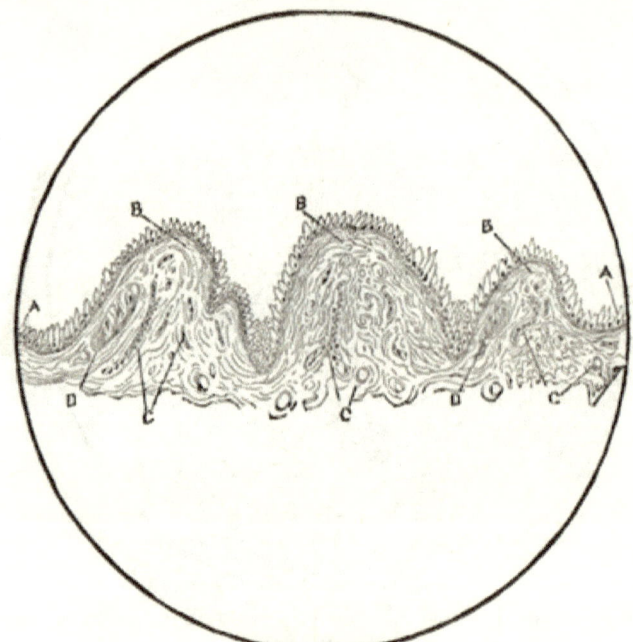

Fig. 54.—Vertical Section of the Derma, or True Skin.

A, A. Line of elongate cells belonging to the epidermis.
B, B, B. Summits of three papillæ of the true skin.
C, C, C. Portions of capillary loops in the papillæ.
D, D. Nerve loops, tactile corpuscles. × 250.

placed vertically to the surface, constitute the *fat-columns of Satterthwaite.*

The blood-vessels supplying the skin may be seen in vertical sections, in the subcutaneous tissue. Branches from these are sent to the papillæ, where they terminate in delicate, interlacing loops of capillaries.

Medullated nerves are also sent to the papillæ; and in certain locations, they may be seen to terminate in tortuous structures—

THE HAIRS.

the *tactile corpuscles*. Varicose nerve fibrils have been traced between the cells in the rete mucosum of the epidermis.

APPENDAGES OF THE SKIN.

The appendages of the skin are the *hairs, sebaceous glands, sudoriferous glands*, and the *nails*.

THE HAIRS.

A hair, consisting of a root and shaft, is constructed from elongate cells which are cemented together, and overlapped with cell-plates. The central part of medullated hairs is composed of cubical cells, pigment, and occasional minute air-bubbles.

The root penetrates the stratum corneum and (appearing to have pushed the rete mucosum before it) passes through the true

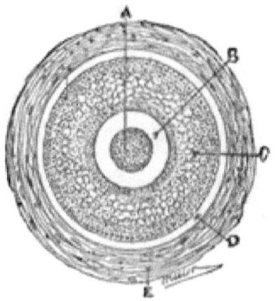

FIG. 55.—TRANSVERSE SECTION OF HAIR, AND HAIR-FOLLICLE. PARTLY DIAGRAMMATIC.

  A. Medulla of hair.
  B. Cortex of same.
  C. Root-sheath.
  D. Glassy membrane.
  E. Fibrous wall of the follicle.

skin and terminates in a bulb usually in the subcutaneous tissue, where it rests upon a papilla composed of an extremely delicate plexus of blood-capillaries.

The Hair-follicle.—The root of the hair, in its passage to the papilla, is invested with sheaths derived from the skin. The hair, with its follicle, is indicated in transverse section in Fig. 55. A represents the *medulla*, and B the *cortex* of the hair. Outside the *root-sheath* C, and derived from the rete mucosum of the epidermis, is a thin layer, the *glassy membrane* D. This is projected from the basement membrane covering the surface of the corium or true skin. The whole is surrounded by a *fibrous coat* E, derived from the connective tissue of the derma.

A vertical section of the follicle is indicated in Fig. 56. A, B, and C represent the epidermal layers which do not enter into its

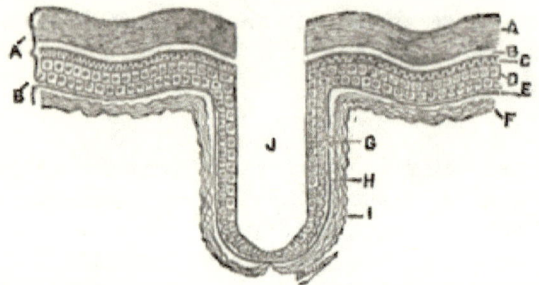

FIG. 56.—DIAGRAM SHOWING MODE OF FORMATION OF HAIR-FOLLICLE.

A'.   Epidermal layers.
B'.   Derma or true skin.
A.   Horny layer of epidermis.
B.   Stratum lucidum.
C.   Stratum granulosum.
The three last mentioned form no part of the follicle.
D.   Rete Malpighii. This will be seen projected into the depths of the true skin to form the root-sheath G.
E.   Hyaline membrane covering the derma. This is projected into the follicle, forming the glassy membrane H.
F.   Fibrous tissue of the derma, forming the fibrous sheath of the hair-follicle I.
G.   Root-sheath of the hair-follicle.
H.   Glassy membrane of the follicle.
I.   Fibrous sheath of the follicle.
J.   The hair-follicle.

composition. The rete mucosum D forms the root-sheath at G. The basement membrane of the corium E forms the glassy membrane H, while the connective tissue F constitutes the fibrous layer of the hair-follicle J.

SUDORIFEROUS GLANDS.

A sweat-gland consists of a tube or duct (*vide* Fig. 57, at A) which, from the opening upon the surface, passes in a spiral course through the several layers of the skin to the deeper part of the corium, where it becomes coiled in a bunch as at D. The coiled or gland part of the tube is surrounded by a network of capillaries. At B, the tube is seen in transverse section. The gland-tube D is provided with a wall of connective tissue and smooth or involuntary muscle, lined with conical cells. The epithelial lining of the duct C is granular; the lumen small and lined with a thin cuticular membrane. The latter constitutes the entire wall of the duct as

SUDORIFEROUS AND SEBACEOUS GLANDS. 73

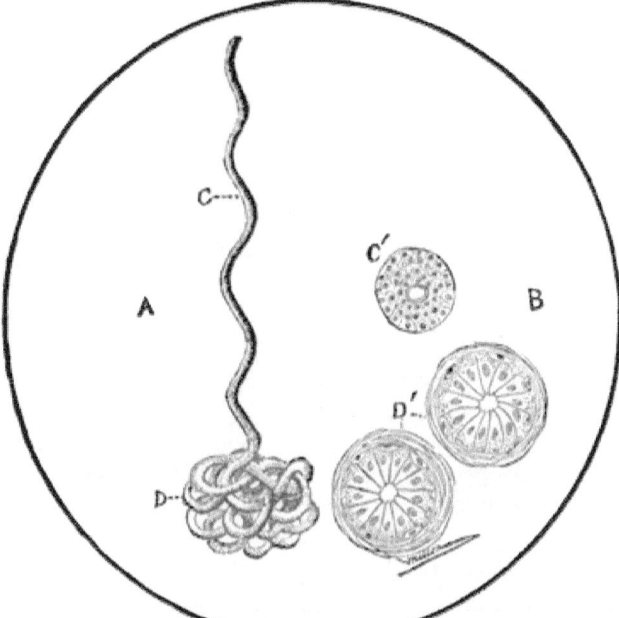

Fig. 57.—SUDORIFEROUS TUBULAR GLAND.

A. Diagrammatic sweat-gland. C. Its duct. D. Coiled, glandular part.
B. The same, showing a transverse section of both parts. × 400. C'. The duct lined with several layers of cells. D'. The coiled glandular part lined with columnar cells in a single layer.

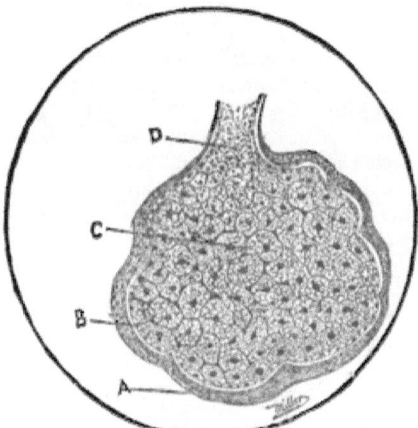

Fig. 58.—SINGLE LOBULE OF A SEBACEOUS GLAND.

A. The fibrous wall of the sac.
B. Involuntary muscular element of the wall.
C. Polyhedral cells filling the sac completely.
D. Fatty degeneration of the parenchyma at the neck of the gland, formation of *sebum*.
× 400.

the surface of the epidermis is approached, the cellular elements having disappeared.

Krause estimated the number of sweat-glands at over two millions.

SEBACEOUS GLANDS.

These glands are little sacs or lobules, one or more of which open into each hair-follicle. These sacs are entirely filled with polyhedral cells (*vide* Fig. 58). At the neck of the gland the cells become granular, fatty, and disintegrated, producing the *sebum*.

MUSCLES OF THE HAIR-FOLLICLES.

Attached to the fibrous layer of each hair-follicle is a small band of involuntary or smooth muscular fibre—the *arrector pili*. This passes obliquely toward the surface of the skin; and when contraction takes place, the follicle and hair are elevated, producing the phenomenon known as goose-flesh.

PRACTICAL DEMONSTRATION.

Remove the skin from the parts below as soon after death as practicable. Tissue may frequently be secured after surgical operations from stumps, etc. Dissect deeply so as to preserve the subcutaneous tissue. Small cubes from the finger-tips, the palm of the hand, the scalp, and the groin may be hardened quickly in strong alcohol; and vertical sections should be made as soon as the tissue has become sufficiently firm. Stain with hæma. and eosin, and mount in dammar.

The structure of hairs may be best demonstrated by washing the soap from lather, after shaving, with several changes of water. When clean, decant the water and add alcohol. After twenty-four hours again decant and add oil of cloves. With a pipette carry a drop of the oil with the deposited hair-cuttings to a slide, remove as much of the oil as possible with slips of blotting-paper, and mount in dammar. Oblique, vertical, and transverse sections may be readily obtained by this method.

VERTICAL SECTION OF SKIN FROM THE GROIN.
(*Vide* Fig. 59.)

OBSERVE:

(L.)*

1. **The horny layer of the epidermis.** (The stratum lucidum will hardly be demonstrable on account of the thinness of the epidermis in this region.)

* Low power—*i.e.*, from thirty to sixty diameters.

2. **The rete mucosum.** (The section from which the illustration has been drawn was taken from a negro, and the deep cells were pigmented.)

3. **The sharp line of demarcation between the epidermis and the true skin.**

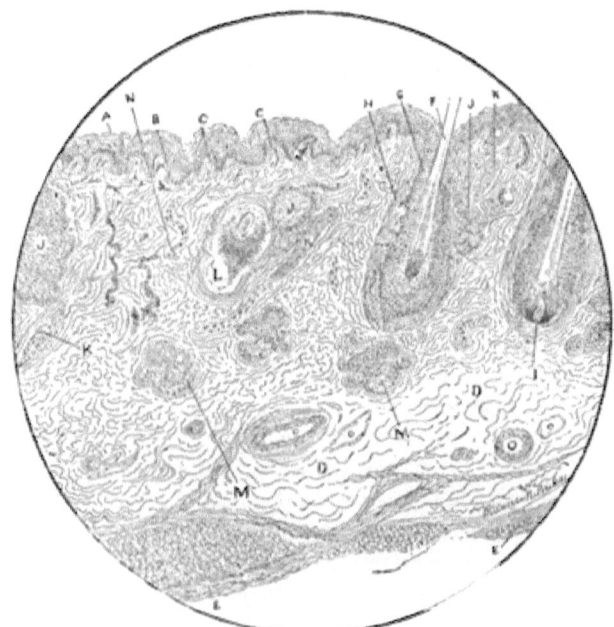

FIG. 59.—VERTICAL SECTION OF SKIN FROM THE GROIN. STAINED WITH HÆMA. AND EOSIN.

A. Epidermis.
B. Deep, elongated cells of the rete mucosum.
C, C. Papillæ of true skin.
D, D. Subcutaneous areolar tissue.
E, E. Collections of adipose tissue.
F. Shaft of hair (obliquely sectioned).
G. Root-sheath of last.
H. Fibrous sheath of last.
I. Hair papilla (vertical section).
J, J, J. Portions of sebaceous glands (one on the extreme right of the cut is seen in connection with the hair-follicle).
K, K. Arrectores pili.
L. Hair-follicle with contained shaft of hair in very oblique section.
M, M. Coils of sudoriferous glands.
N. Spiral duct of last.
O, O. Arteries of subcutaneous plane.

4. **The papillæ of the corium** or derma. (Note the absence of any sharp line dividing the corium and subcutaneous tissues.)

5. **The larger blood-vessels** of the subcutaneous region.

(The arteries in transverse section are plainly indicated by their prominent media, the appearance of the fenestrated membrane as a wavy yellowish line, and by the elliptical or circular outline. The **veins** are smaller, with thinner walls, and their outline is generally irregular. The smaller veins are commonly overlooked, on account of their lumen having become obliterated by contraction of the tissue in hardening.)

6. **Coils and ducts of sweat-glands** in last region. (The tubes are cut in various directions, and the whole is surrounded by dense fibrous tissue, forming a kind of capsule.)

7. The **collections of adipose tissue** beneath the last region. (The septa are dense and strong.)

8. (Having selected a vertical section of a hair-follicle:) (*a*) The **root of the contained hair.** (*b*) The **bulb and the hair papilla.** (*c*) The **medulla of the hair.** (*d*) The **root-sheath** prolonged from the rete mucosum. (*e*) The **fibrous** (outer) **sheath.**

9. The **sebaceous glands.** (The demonstration of the connection between the neck of the gland and the follicle will require a very favorable section.)

10. (Scattered through the corium and upper subcutaneous region:) (*a*) Small **portions of sebaceous glands.** (*b*) **Ducts of sudoriferous glands.** (*c*) **Oblique sections** at various angles of **hair-follicles.** (*d*) **Small vessels.**

11. **Arrector pili muscle.** (Nearly always to be found standing obliquely to the divided hair-follicle.)

(H.)*

12. (If demonstrable:) (*a*) The **stratum lucidum.** (*b*) **Stratum granulosum.**

13. The **elongate cells of the rete,** next the corium.

14. (Where the tissue has been torn:) The **impacted cells** of the horny epidermis.

15. The basement **membrane covering** the corium.

16. **Capillaries of the papillæ of the corium.** (These may be distinguished, when seen longitudinally, by tortuous lines of elongate and deeply-stained nuclei belonging to the endothelium. Arterioles may be differentiated by their long muscle cells, the circular fibres lying transversely to the vessel.)

17. The **root-sheath of the hair-follicles.** (The cells composing the root-sheath vary in appearance, according to their posi-

---

* High power—*i.e.*, from three to four hundred diameters.

tion relatively to the hair; and this will enable you to demonstrate two layers, or an inner and an outer root-sheath.)

18. The **glassy membrane of the hair-follicle.** (Appearing simply as a clear space between the root-sheaths and the outer fibrous coat.)

19. The **intra-cellular network** in the large polyhedral cells of the sebaceous glands, and the minute **fat-globules** in the same.

20. The **nuclei of the fat-cells** in the adipose tissue. (They appear pressed to one side.)

21. **Medullated nerve bundles** in transverse or oblique section.

THE TEETH.

A human dentinal tooth is a calcific structure of extreme hardness, and is divided into an exposed *crown*, a constricted *neck*, and one or more concealed *fangs*—the latter being inserted into an alveolus, by means of which the whole is firmly connected with the maxilla.

The central portion presents an elongate cavity (*pulp-chamber*) containing vascular, nervous, and connective-tissue elements—the *pulp*.

The pulp-cavity is surrounded by the *dentine*, which constitutes the major portion of the tooth.

The crown portion of the dentine is provided with a covering of *enamel*, while the fang is invested with an osseous cement, the *crusta petrosa*.

A thin (one-twenty-five-thousandth to one-fifty-thousandth of an inch) membrane—the *cuticula*—covers the enamel in early life, while the crusta receives a periosteal investiture. The vascular and nervous elements of the pulp obtain admission to the pulp-cavity by a perforation or foramen at the apex of the fang, the *foramen dentium*.

The Pulp.—The ground-substance, or stroma of the pulp, is a form of primitive connective tissue, gelatinous rather than markedly fibrous. It contains elongate capillary loops, multipolar cells medullated and non-medullated terminal nerve fibrils.

Surrounding the pulp mass, and next to the dentinal wall of the chamber, we find a single layer of elongate cells—*odontoblasts*. These are in communication, by means of processes, or prolongations, with fibrous elements of the pulp.

Dentine.—The dentinal stroma or matrix is cartilaginous, with calcific elements, and is, next to the enamel, the hardest tissue of the body. The matrix is pierced with the *dentinal canals* (one-ten-thousandth to one-twenty-thousandth of an inch in diameter), which radiate from their beginning, next the pulp-chamber, toward the outer portion of the dentine. These canals branch and anastomose, and are lined with an exceedingly thin *dentinal sheath*.

From the outer extremity of the odontoblasts of the pulp numerous prolongations are sent which are continued within the dentinal canals as the *dentinal fibres*. The dentinal canals terminate exteriorly, by very fine lumina, in a system of irregularly-formed openings, *interglobular spaces*, which are channelled in the outer part of the dentine. The dentinal terminal fibres are in connection with branched cells which occupy the interglobular spaces.

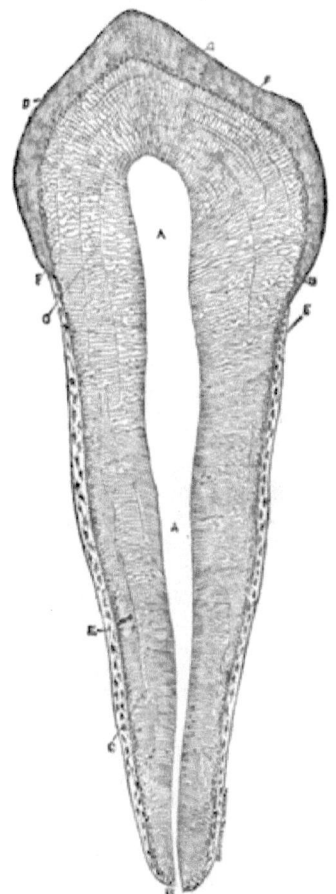

FIG. 60.—VERTICAL SECTION OF BICUSPID TOOTH. DIAGRAMMATIC.

A, A. Pulp-chamber.
B. Foramen dentium for entrance of vascular and nervous elements to the pulp.
C, C, C. Dentine. The lines point to the incremental lines of Salter, imperfectly calcified dentinal stroma.
D, D. Interglobular spaces in last layer, forming the granular layer of Purkinje.
E, E. Crusta petrosa or cement substance.
F. Enamel. The parallel lines are intended to indicate the stripes of Retzius due to the formation of the enamel in successive layers.
G. The cuticula.

The Enamel.—The part of the dentine above the neck of the tooth is protected by a covering of *enamel*. The enamel consists of *prisms* from one-six-thousandth to one-eight-thousandth of an inch in diameter which pass in a direction nearly at right angles to the surface of the dentine. They are of extreme density, contain little besides inorganic material, and in a vertical section the whole is traversed by parallel striæ, not unlike the markings indicating tree-growth—the *lines of Retzius.*

Crusta Petrosa.—The fang portion of the dentine is invested with a thin layer of true bone, arranged in *laminæ* and containing *lacunæ* and canaliculi, but no Haversian canals. The crusta is provided with *periosteum*, which forms the bond of union between the teeth and the process of the maxillæ. The lacunar bone corpuscles are in connection, through the canaliculi, with the cells in the interglobular spaces of the dentine. It will be seen that the connective-tissue elements, at least of the pulp, are in eventual histological connection with the bone corpuscles of the crusta.

PRACTICAL DEMONSTRATION.

The illustrations common to our text-books have been drawn from dried teeth, ground down to the requisite thinness by means of corundum or emery wheels. This is a very tedious process, and is impracticable with the student. If such specimens are desired it will be advisable to purchase them already mounted. They only give the skeleton of the organ, all the soft tissues being destroyed by the drying and grinding.

While dry specimens exhibit the plan of a tooth, the soft tissues must be studied in sections made after the inorganic constituents have been removed. Teeth immediately after extraction are to be treated in the same manner as described for bone. A one-sixth per cent of chromic-acid solution, to which five drops of nitric or hydrochloric acid have been added, may be first used. Let the quantity of liquid be liberal, and from time to time, say every three days, add a few drops of the nitric acid. The decalcification should proceed slowly and may be complete in from two to three or four weeks. The earthy matters will first be dissolved from the surface. Watch the action carefully, ascertaining the progress of decalcification by pricking a fang with the needle. If the acid be too strong, and the action too rapid, the whole may be destroyed. When the decalcification is complete, a needle may be easily passed through the tooth and sections be made with the razor or knife, with or without a microtome. The form will be preserved except as regards the enamel; this will be entirely dissolved. The enamel prisms may be demonstrated by treating broken fragments with dilute acid for a short time only.

Sections should be stained with carmine and picric acid and mounted in glycerin. For the study of the development of teeth, fœtal jaws may be treated as just described; and, when properly decalcified and hardened, should be infiltrated with celloidin, sectioned, and stained. I would refer the student to the excellent article on the subject in Dr. C. Heitzmann's "Morphology."

TRANSVERSE SECTION OF FANG OF HUMAN DECIDUOUS CANINE TOOTH.—DECALCIFIED.
(Fig. 61.)

OBSERVE:

(L.)

1. Division into pulp, dentine, crusta petrosa, and periosteum.

2. Line of **junction of pulp and dentine.** (If the elements of the pulp are intact, note the **layer of** deeply-stained **odontoblasts** next the dentine.)

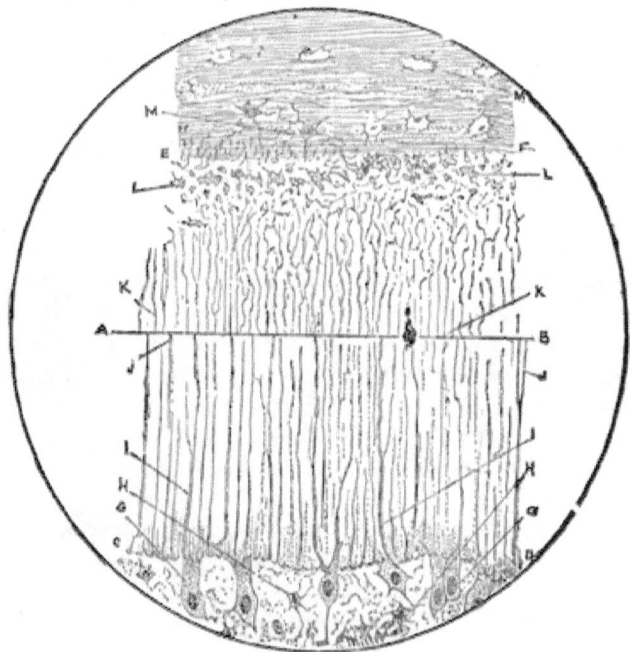

FIG. 61.—TRANSVERSE SECTION OF FANG OF A DECIDUOUS CANINE TOOTH, DECALCIFIED WITH CHROMIC AND NITRIC ACIDS AND STAINED WITH PICRO-CARMINE.

A, B. Line through the dentine indicating the point at which the edges have been made to join after the omission of an intervening portion. This was necessary in order that the different layers might be shown in a single drawing.

C, D. Junction line between the pulp and dentine.
E, F. Junction line between dentine and crusta petrosa.
G, G. Odontoblasts of the pulp.
H, H. Stellate connective-tissue cells of the pulp.
I, I. Dentinal processes of odontoblasts.
J, J. Dentinal fibres.
K, K. Terminal branching dentinal fibres.
L, L. Interglobular spaces of dentine.
M, M. Lacunæ of the crusta petrosa. The drawing does not show the periosteal investiture of the crusta. × 400.

3. **External limit of dentine.** (Note here the deeply-stained granular **line of Purkinje.** This is the location of the interglobular spaces. The deep color is due to the staining of their cell contents.)

4. The **striæ of the dentine** (dentinal canals and stained contents).

5. The **laminated crusta.** (The yellowish-pink dots on the lacunæ.)

(H.)

6. **Elements of the pulp.** (*a*) The layer of **odontoblasts** (note their **internal processes** connecting with other cells of the pulp; and the **external processes** passing into the dentinal canals). (*b*) The **sparsely fibrillated character of the pulp tissue.** (*c*) Sections of **vascular loops.** (The **nerve elements** may be demonstrated, particularly if the section be made near the apex of the fang, where the fibres are medullated. The terminal fibrillæ are non-medullated.)

7. **Dentinal elements.** (*a*) The **dentinal canals.** (*b*) The **dentinal sheath.** (Better demonstrated in transverse sections.) (*c*) **Dentinal fibres.** (In transverse sections the canals are well shown lined with a membrane of extraordinary tenuity, with the fibre appearing as a central dot.) (*d*) **Fine dentinal fibres** near the outer limit. (*e*) **Interglobular spaces.** (An occasional cell may be made out in the larger spaces. They were formerly supposed to contain a gelatinous material only. Note the connection between these spaces and the termini of the dentinal fibres.)

8. The **Crusta Petrosa.** (*a*) Its **laminated formation.** (*b*) The **lacunæ.** (*c*) **Bone corpuscles** in the last. (The canaliculi are not well demonstrated here, as the tissue is very translucent and feebly stained. These minute canals are better indicated in dried bone.)

9. **The periosteum.** (Note its dense fibrillar meshwork.)

THE STOMACH AND INTESTINES.

The stomach and intestines are lined with mucous membrane, i.e., a membrane containing glands which secrete mucus.

The gastric and intestinal mucous membranes are constructed as follows:

1. *The epithelial lining.*
2. *The mucosa.*
3. *The muscularis mucosæ.*
4. *The submucosa.*
5. *The muscular walls proper.*
6. *The fibrous or peritoneal investment.*

In descriptive anatomy, the first four of the above are included in the mucous coat.

The epithelium of the inner surface of that portion of the alimentary tract under consideration is of the columnar variety. Variations occur in the deeper layers, which will be referred to later on.

The mucosa, with its epithelial covering, is thrown into coarse folds, *rugæ* or valve-like reduplications, which greatly increase the extent of surface. It contains the principal glands and capillary blood-vessels.*

The muscularis mucosæ is a thin layer of involuntary muscular fibre which separates the mucosa from the submucosa.

The submucosa, composed of loose areolar tissue, serves to connect the previous structures with the muscular coat proper, and contains the larger trunks from which the capillaries of the mucosa either take their origin, or into which they empty. An intricate plexus of lymphatics is also here situated.

The muscular coat consists of strong bands, running in two or three directions. The muscle plates are sustained by connective tissue.

A peritoneal investment covers the organs, except at such points as are occupied by the entrance and exit of blood-vessels, etc.

THE STOMACH.

The mucosa everywhere contains microscopical depressions, the *gastric tubules* or *peptic glands*. These are concerned in the production of the gastric juice, and in the absorption of fluids.

---

* It is not always possible in mucous membranes to differentiate clearly between an epithelial lining and the mucosa; and in the stomach and intestine they may be both included in the mucosa.

The several layers of the stomach may be better understood by reference to the diagram (Fig. 62).

The gastric tubular glands are of two principal varieties, viz.: 1, the *peptic glands*, found in the cardiac portion of the stomach; 2, the *pyloric glands*, which occupy the pyloric extremity of the organ. The mucous membrane, midway between the cardiac and pyloric portions, is occupied by tubules which partake of the character of both peptic and pyloric glands, so that no sharp boundary line exists.

The peptic or cardiac gland-tubes penetrate to the muscularis

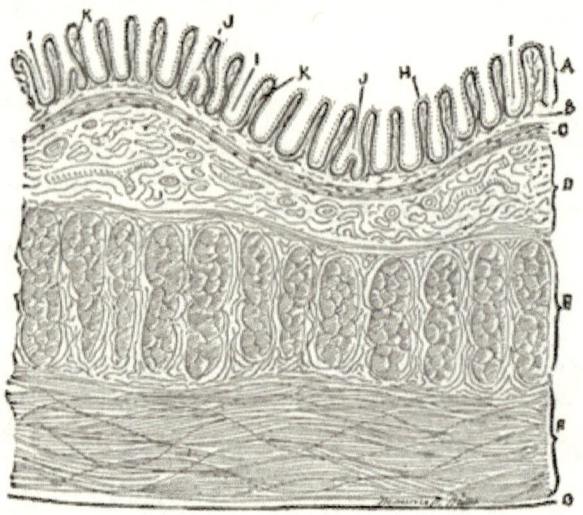

Fig. 62.—Diagram of the Wall of the Stomach in Vertical Section.
A. Layer of gastric tubules.
B. Vascular portion of mucosa.
C. Muscularis mucosæ.
D. Submucosa.
E. Internal circular layer of muscular fibre.
F. External oblique and ongitudinal muscular layers.
G. Peritoneum.
I, I, I. Lumen of gastric tubules.
J, J. Branching gastric tubules.
K, K. Blood-vessels arising from lower portion of mucosa, forming plexus between the tubules.

mucosæ. They pursue a somewhat wavy course, and at their lower or blind extremity are frequently bifid. They are lined at their commencement on the surface with translucent columnar epithelium, the cells being polygonal in transverse section. As the fundus or bottom of the tube is approached, the lining cells become granular, larger, and somewhat polyhedral. Next the wall of the tube, large, granular, bulging cells are scattered irregularly. The epithelium occupies the major portion of the space in the tube, so that the lumen is very small.

A single bifid tube is represented in Fig. 63. The prominent distinguishing feature of the peptic or cardiac tubules is afforded by the large *border* or *parietal cells*. The cells next the lumina are called *central* or *chief cells*.

The pyloric gland-tubes pursue a course not greatly unlike that of the tubes just mentioned. They do not branch, however, until they have penetrated well down toward the muscularis mucosæ.

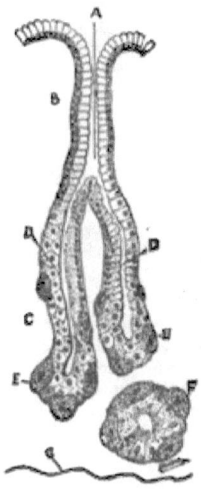

FIG. 63.—VERTICAL SECTION OF A PEPTIC TUBULAR GLAND, FROM CARDIAC MUCOSA OF STOMACH. LARGELY DIAGRAMMATIC.

    A. Lumen of duct portion of tubule.
    B. Neck of last.
    C. Gland portion.
    D, D. Central cells.
    E, E. Border cells.
    F. The glandular portion in T. S.
    G. Line of commencing muscularis mucosæ.

Their distinguishing character is afforded by the epithelial lining. At the surface, the cells are columnar with polygonal transection. The deeper parts are lined with translucent cylinders. The lumina are larger than those of the peptic tubes.

The gastric gland-tubes are placed thickly side by side, their bases reaching the muscularis mucosæ. Between and beneath the tubes is a dense network of blood capillaries.

The remainder of the stomach has little special interest for the histologist. The muscular portion of its walls consists of a thin internal circular layer, with oblique bundles interspersed, and a thin external longitudinal layer of the involuntary variety. Between the two layers is found a plexus of non-medullated nerves,

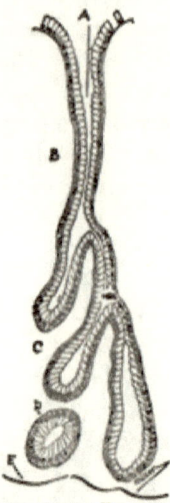

FIG. 64.—VERTICAL SECTION OF TORTUOUS AND BRANCHING TUBULAR GLAND, FROM PYLORIC MUCOSA OF STOMACH. DIAGRAMMATIC.

    A. Lumen. This is often much widened.
    B. Duct portion of tubule.
    C. Branching glandular portion.
    D. Transverse section of the last.
    E. Lower limit of mucosa.

corresponding to the plexus of Auerbach of the intestines, but which is not demonstrable by ordinary methods or sections.

The blood-supply is received at the curvatures. Branches penetrate the muscular layers along the lines of omental attachment, as blood-vessels never penetrate the peritoneum.

The peritoneum is constructed mainly of fibrous tissue, with an external investment of pavement epithelium.

PRACTICAL DEMONSTRATION.

Inasmuch as the human stomach cannot often be obtained until decomposition has destroyed it for our work, we must secure the organ from some one of the lower animals. The stomach of the dog presents all the histological features of that of man, and can be gotten in good condition from an animal recently killed.

Harden small pieces in strong alcohol, and cut sections at right angles to the surface and from different regions. Stain with hæma. and eosin, and mount in dammar.

VERTICAL SECTION FROM GREATER CURVATURE OF DOG'S STOMACH.

(Fig. 65.)

OBSERVE:

(**L.**)

1. The division into: (*a*) **Surface epithelium** (free ends of gland-tubes). (*b*) **Mucosa.** (*c*) **Muscularis mucosæ.** (*d*) Sub-

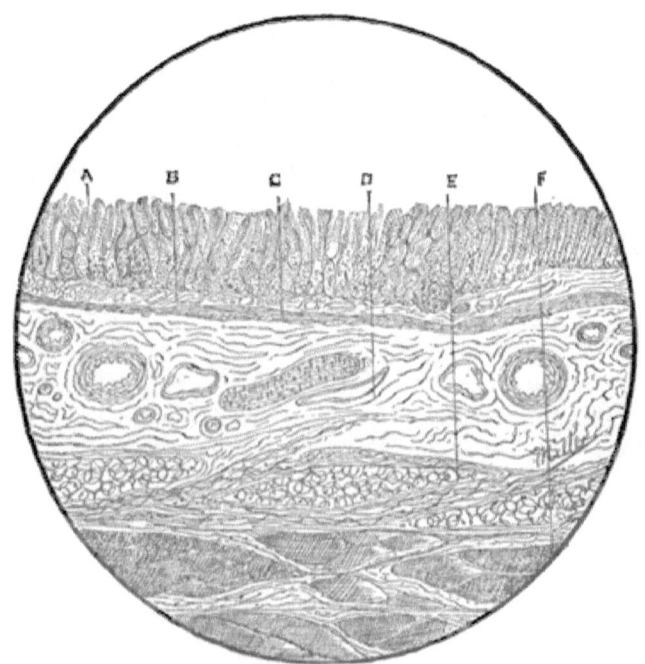

FIG. 65.—VERTICAL SECTION OF WALL OF CENTRAL PORTION OF DOG'S STOMACH.

A. Internal surface, showing open mouths of the gastric tubules, lined with clear columnar cells.
B. Deepest portion of submucosa.   C. Muscularis mucosæ.
D. Submucosa.   E. Adipose tissue in last.
F. Bundles of muscular tissue (internal circular).   × 60.

mucosa. (*e*) **Muscular layers.** (Only a portion of the inner circular layer is shown. It has been divided transversely.)

(**H.**)

2. The **epithelium of gland-tubes.** (The upper portion of the tubes will be cut obliquely in many places, as they have been inclined, and the epithelium will show as a beautiful mosaic of polygonal areas.) (*a*) The differentiation between **border and**

central cells. (*b*) **Tubes cut transversely,** showing the lumina. (*c*) Indications of the **capillary plexuses between the tubes.**

3. The **mucosa.** (*a*) **Arterioles and venules** beneath the tubules. (*b*) Scattered **lymphoid cells** (round cells with one, two, or three nuclei).

4. The **muscularis mucosæ.** (Note the elongated nuclei of the smooth muscle cells.)

5. The **submucosa.** (*a*) **Arteries, veins, etc.,** cut in various directions. (*b*) The **adipose tissue.** (Crystals of the fatty acids are frequently seen in the cells when freshly mounted.)

6. The **muscular bundles of the circular layer** with the **septa of connective tissue.** (Note particularly the various appearances presented by bundles of involuntary muscular fibre when cut in different planes.)

SMALL INTESTINE.

The histology of the intestine, both large and small, is formed upon the general plan of that of the stomach. The same layers are presented: the *mucosa*, with its epithelial covering; the *muscularis mucosæ;* the *submucosa;* the *muscular* and the *peritoneal* coats.

The mucosa of the small intestine is everywhere pierced by blind depressions; or, what is equivalent, the surface is studded with minute elevations or papillæ, between which are the *depressions which correspond to the tubules of the stomach.* The elevations are called *villi,* the depressions between the villi, *crypts.*

The small intestine serves two important functions: 1. The *secretion of a fluid,* one of the digestive juices—the *succus entericus.* 2. The *absorption of food,* especially the fats or hydrocarbons.

We shall view the histology of this organ from a physiological standpoint, considering: 1st, *Those structures concerned in the secretion of the succus entericus;* 2d, *Those portions concerned in absorption of food.*

HISTOLOGY OF THOSE PARTS OF THE SMALL INTESTINE PARTICULARLY CONCERNED IN THE PRODUCTION OF THE SUCCUS ENTERICUS.

The diagram (Fig. 66) is intended to represent at A the thickness of the mucosa with its papillary elevations—the villi. The muscularis mucosæ B, from which the villi arise, separates the mucosa from the submucosa C. The horizontal line at the bottom of the diagram indicates the outer limit of C and the beginning of the circular muscular coat of the intestine. The *villi*, everywhere covered with columnar epithelium, are represented in the drawing as widely separated, but in the gut they are so closely studded as to afford but narrow chinks (*crypts*) between the prominences. In the interior of each villus is a fine network of *blood capillaries* (G G). The cells on the borders of the villi secrete certain fluid material from the blood circulating in the capillary plexuses, and pour it out into the crypts. The crypts becoming filled with the fluid, the latter overflows and passes into the lumen of the gut, to act in promoting digestion. This is one source of the succus entericus, and there is yet another.

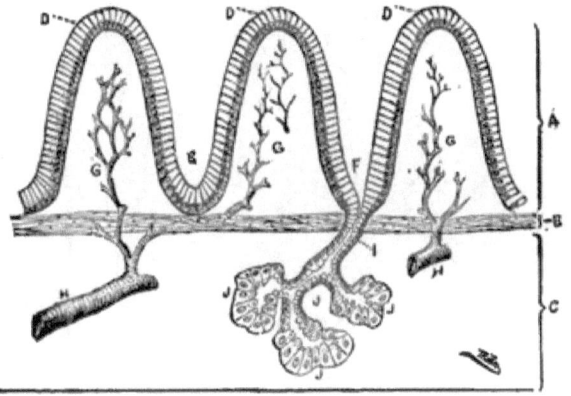

Fig. 66.—Diagram showing Portions of Intestinal Mucous Membrane, concerned in the Secretion of the Succus Entericus.

A. The mucosa.
B. Muscularis mucosæ.
C. Submucosa.
D, D, D. Villi.
E, F. Crypts of Lieberkühn.
G, G, G. Blood plexuses of villi.
H, H. Large vessels of submucosa, supplying the epithelium covering the villi.
I. Neck of a gland of Brunner.
J, J, J. Gland of Brunner in the submucosa. The secretion is emptied into the crypts as at F.

From the bottom of some of the crypts, tubes will be found which, piercing the muscularis mucosæ, reach the submucosa, where they branch, become convoluted, are lined with secreting cells, and are known as the *glands of Brunner*. These glands, which are practically elongated crypts, are surrounded by blood capillaries, and the gland-cells secrete a fluid which is poured into the gut at the base of the crypts, when it becomes mingled with the secretion previously mentioned, and constitutes the *succus entericus*.

We have, then, seen that the succus entericus is secreted, *partly from the epithelial cells covering the villi* (or, in other words, surrounding the crypts) and *partly from the cells of Brunner's glands*.

THE REMAINING STRUCTURES OF THE INTESTINE CONCERNED MAINLY IN FOOD ABSORPTION.

The diagram (Fig. 67) is intended to show the same layers as were indicated in the previous figure (*Brunner's glands and the blood-vessels have been omitted* in order to avoid confusion). The villi and crypts are seen as before.

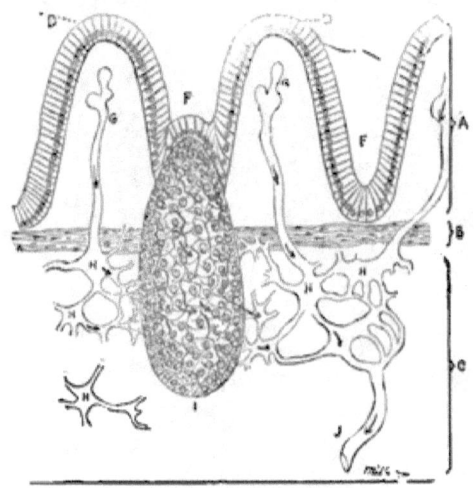

FIG. 67.—DIAGRAM SHOWING PORTIONS OF INTESTINAL MUCOUS MEMBRANE, CONCERNED IN ABSORPTION.

A. Mucosa.
B. Muscularis mucosæ.
C. Submucosa.
D, D. Villi.
E, F. Crypts of Lieberkühn.
G, G. Lacteals.
H, H. Chinks and intercommunicating channels of the lymph plexus of the submucosa.
I. Bottom of a mass of adenoid tissue—a so-called solitary gland. Peyer's patches are formed of aggregations of these nodules.
J. Efferent lacteal or lymph duct.

In the centre of each villus is the blind tube G G, a part of the lymphatic system, and here called a *lacteal*. When, during digestion, the minute globules of fatty food reach the small intestine, they are grasped by the epithelial cells covering the villi, and are carried eventually within the body of the villus to this lacteal.

The lacteals pierce the muscularis mucosæ, and in the submucosa are in connection with a *plexus of lymphatic tubes and spaces*. They eventually unite with *efferent lymph-tubes* (J), and pass by means of the mesentery to the *receptaculum chyli*.

Connected with the plexus of lymphatics in the submucosa are minute *nodules of lymphoid structure* (adenoid tissue), which have unfortunately been called lymphatic *glands*. They are in no sense glands.

Slit up a portion of intestine along the attached border, and carefully examine the inner surface: it will present a velvety appearance, due to the minute villi. You will also find little nodules, perhaps one-sixteenth of an inch in diameter, scattered here and there in the mucous coat. These are the lymphatic nodules alluded to above—the so-called *solitary glands*. One of the nodules is indicated in the diagram at I, with its point projecting into the crypt F.

Continuing your examination of the gut, you will discover, especially in the ileum, roughened patches perhaps two inches long by half an inch broad. These are collections of the lymphatic nodules described in the last paragraph, and are termed *agminate glands* or *patches of Peyer*. They have no secretive power, being simply in connection with, and a part of, the chain of lymphatics in the walls of the intestine. They consist of adenoid tissue, which will be described with the lymphatics.

To recapitulate, the small intestine presents the following:

1. The *villi*, each containing a plexus of blood-capillaries and the lymphatic or absorbent vessel.
2. *Crypts* or follicles *of Lieberkühn*, which are simply depressions between the villi.
3. *Brunner's glands*, the only true glands of the gut, unless the crypts are so classified.
4. *Solitary lymphatic nodules*, the so-called solitary glands.
5. *Agminate lymphatic nodes*, agminate glands or patches of Peyer, consisting of aggregations of solitary lymphatic nodules.

The muscular part of the intestine is arranged not unlike that portion of the stomach, *i.e.*, with an inner circular and an outer longitudinal layer. Between the two is located *Auerbach's plexus* of non-medullated nerves. A similar plexus, *Meissner's*, is found in the submucosa. These we shall not attempt to demonstrate.

A small quantity of areolar tissue connects the external longitudinal muscular layer with the peritoneal investment.

PRACTICAL DEMONSTRATION.

The intestines of the dog or rabbit are more commonly used for practical work, for reasons already alluded to. The tissue should be cut in small pieces, and hardened quickly in alcohol. When human intestine can be obtained fresh, a piece, say three inches long, should be emptied of its contents, filled with alcohol by tying the ends, and the whole hardened in strong spirit. Under no circumstances should the gut be washed, and great care must be taken to avoid injuring the delicate cells covering the villi. Vertical sections with the microtome are the most valuable. Stain with hæma. and eosin, and mount permanently in dammar.

VERTICAL SECTION OF THE ILEUM, INCLUDING PORTION OF A PATCH OF PEYER. HUMAN.

(Vide Fig. 68.)

OBSERVE:

(L.)

1. The villi. (*a*) **That they are of varying lengths, slender, wavy, and delicate.** (*b*) **The covering of columnar**

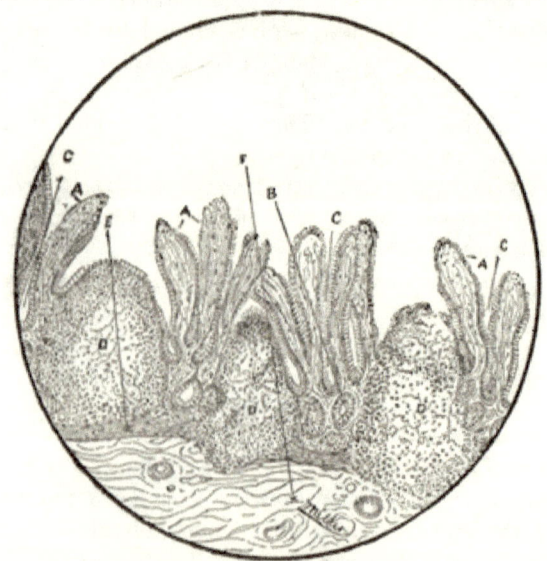

FIG. 68.—INTESTINAL MUCOUS MEMBRANE THROUGH A PEYER'S PATCH, VERTICAL SECTION. Stained with Hæma. and Eosin. × 250.

A, A, A. Villi.
B. Transverse sections of crypts of Lieberkühn.
C, C. Crypts in vertical section.
D, D, D. Nodules of lymphoid tissue—constituting a patch of Peyer.
E. Muscularis mucosæ.
F. Submucosa.

cells. (The free extremities of many of the villi in the drawing are seen broken, and the epithelium is wanting in places. It is almost impossible to secure perfect villi from human intestine, on account of the length of time usually intervening between death and the removal of the tissue.) (*c*) **Oblique sections.**

2. The **crypts of Lieberkuhn.**

3. The **lymphatic nodules** (so-called solitary glands), constituting the elements of a patch of Peyer. (*a*) **Their projection upon the mucous surface of the gut between the villi.** (*b*) **The covering with epithelium** on their free borders. (They are located, properly speaking, in the submucosa and between the villi. In the drawing, their bases do not all appear in the submucosa, inasmuch as the nodules are cut in different planes.)

4. **Muscularis mucosæ.** (*a*) The **elongate nuclei** of the involuntary muscular element.

5. The **submucosa.** (*a*) The **blood-vessels.** (*b*) **Lymph-spaces.** (Lymphatic channels are very irregular in form and size, and are often mistaken, in sections, for ruptures in the connective tissue. The stained nuclei of the endothelial cells, with which all lymph channels are lined, will enable you to differentiate.) (*c*) **Glands of Brunner.** (There are none shown in this section. The glands consist of convoluted, branching tubes which penetrate from the crypts to the submucosa. They are lined with columnar epithelium, and as they are divided in a section, they resemble very nearly a crypt of Lieberkühn. Extensive groups are found in the duodenum at its pyloric origin.)

(H.)

6. **The villi.** (*a*) **The covering columnar cells.** (*b*) **Beaker cells** scattered between the last. (These beaker, goblet, or mucous cells are well shown in the intestine of the dog or rabbit.) (*c*) **The lacteals.** (These are not plainly demonstrable, under ordinary circumstances, in human tissue. Sections from the gut of a dog killed during the active digestion of materials rich in hydrocarbons, will show them filled with minute fat-globules.) (*d*) The **basis tissue,** a fibrous reticulum containing many lymphoid cells. (*e*) **Portions of the capillary plexuses.**

7. **Blood-vessels of the mucosa** below the villi.

8. The **adenoid tissue** of the lymph nodules.

THE LUNG.

BRONCHIAL TUBES.

At the root of each lung the large primary bronchus enters, and immediately divides into two equal branches—dichotomously. It is evident that if this mode of subdivision were continued, the

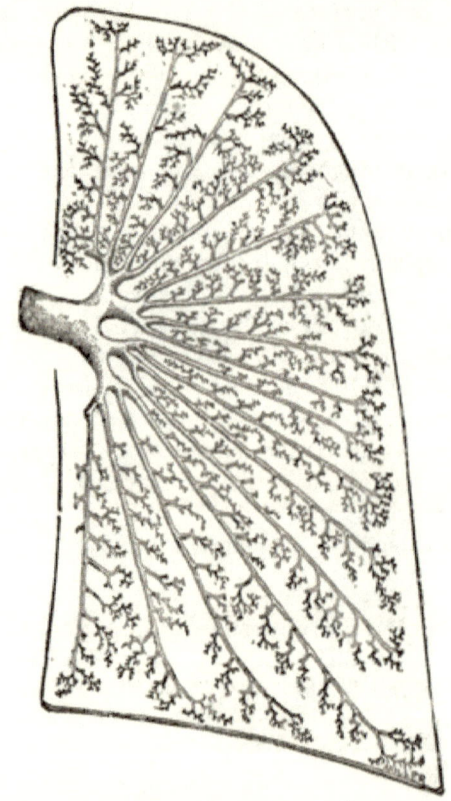

FIG. 69.—DIAGRAM SHOWING THE PLAN OF SUBDIVISION OF BRONCHI, IN THE HUMAN LUNG.

As the main bronchus enters the organ it is seen to divide, dichotomously, until the resultant branches become quite small—say one-tenth inch. These small bronchi now pursue a straight course toward the periphery of the lung, at the same time giving off branches spirally. The last divide dichotomously and result in the terminal, ultimate, or capillary bronchi.

periphery of the organ alone would contain minute bronchi. The arrangement is, however, such as to give everywhere throughout the lung, bronchial twigs, terminal or capillary bronchi, from one-

one-hundredth to one-two-hundredth of an inch in diameter, as follows:

The dichotomous subdivision is continued until the resulting branches become reduced to about one-sixth of an inch in diameter, when *this mode of division ceases*, and the resulting *tubes are projected radially* toward the periphery of the lung. As the straight tubes pursue their course, side branches are given off in spiral succession. The side tubes themselves give off branches which divide dichotomously into the terminal bronchi. The straight tubes constantly diminish in size, and ultimately divide and result also in terminal bronchi. The diagram (Fig. 69) is intended to illustrate this plan of subdivision, but it is purely schematic.

A typical bronchial tube (Fig. 71) presents four coats as follows:
1. *Epithelial.*
2. *Internal fibrous or mucosa.*
3. *Muscular or muscularis mucosæ.*
4. *External fibrous or submucosa.*

The *lining epithelium* is composed of cylindrical cells, provided on their free extremities with delicate hair-like appendages—the *cilia*. Between the pointed, attached end of the ciliated cells, small ovoid cells are wedged, and the whole rests upon a layer of round cells. The epithelium pursues a wavy course, so that the lumen of a tube appears stellate rather than circular in transverse section. This greatly increases the extent of surface.

The *internal fibrous coat* or mucosa is composed of a small amount of connective tissue, which, just beneath or outside the epithelium, sustains collections of *adenoid* or *lymphatic tissue*. In the pig, a considerable quantity of *yellow elastic tissue* is found in the mucosa outside the adenoid tissue, but the amount is smaller in man. The fibres are for the most part disposed longitudinally. Many *nutrient vessels* from the bronchial artery, capillaries, venules, and lymph-spaces, are also found in this coat.

The *muscular coat*—muscularis mucosæ—does not differ from the same layer in other mucous membranes. Its thickness varies in proportion to the size of the bronchus, the smaller tube possessing relatively the thicker walls. The fibres pass circularly, and are of the non-striated or involuntary variety.

The *external coat* or submocosa is largely composed of loose connective tissue, the fibres being mostly arranged circularly. A few delicate elastic fibres run longitudinally. The external fibres, like those of all tubes, ducts, and vessels, are for the purpose of establishing connection with the organ or part traversed; so that it is often difficult to demonstrate the exact external limit of a bronchus. This coat is liberally supplied with nutrient branches from the bronchial artery.

The elasticity and strength of the larger and medium-sized bronchi are greatly increased by the presence of *cartilage* in the form of *plates*, which are imbedded in the external coat. They are not uniform in size, neither are they placed regularly. They frequently overlap one another, and two or three may be superposed. As the tubes become reduced in size the plates become diminished in frequency—disappearing altogether when a diameter of about one-twentieth of an inch has been reached. The cartilage is of the hyaline variety; and each plate is covered with a dense fibrous coat, the *perichondrium*, which unites it with contiguous parts.

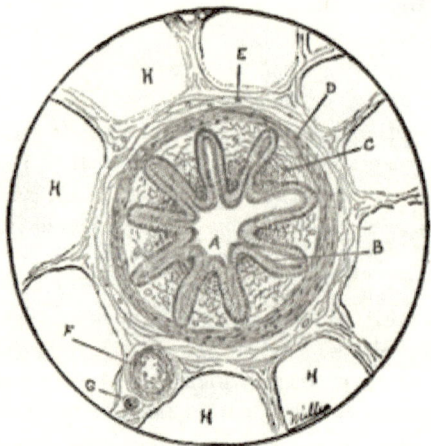

Fig. 70.—Transverse Section of a portion of Human Lung, showing a small Bronchus. Stained with Hæma.
- A. Lumen of bronchus.
- B. Ciliated columnar epithelium.
- C. Internal fibrous layer—*Mucosa*.
- D. Muscular coat.
- E. External fibrous layer—*Submucosa*.
- F. Pulmonary artery.
- G. Nerve.
- H, H, H. Pulmonary alveoli surrounding bronchus. × 60.

The principal bronchi are provided with a great number of *mucous glands*, which are located in the external coat or submucosa. They are simple, coiled tubular glands; commencing on the inner surface, penetrating the mucosa and muscularis mucosæ, and terminating in the submucosa, generally within the cartilage, where they are coiled in short, close turns resembling, in sections, somewhat the larger sweat-glands of the skin. The ciliated epithelium of the bronchus is continued down the beginning of the tube for a short distance, after which the cells are shortened, and lose their cilia. The coiled, gland-part of the tube is lined with conical cells, which are so large as to leave the lumen very small. Sometimes, and especially in the aged, an ampulliform dilatation of the tube may be seen during its passage through the mucosa.

The description just given will apply to large and medium-sized bronchi. Very important changes take place as we pass to the terminal tubes.

As the tubes decrease in size, the first coat to diminish in thickness is the outer, or submucosa. We have already alluded to the disappearance of the cartilage, and the mucous glands are lost at about the same time. The outer coat becomes, in the small bronchi, so thin as to be no longer distinctly demonstrable. The muscular coat is the last to disappear. It remains a prominent feature of the tube as long as separate coats can be distinguished. The epithelial cells lining the tubes toward the termini become shortened, and, getting lower and lower, at last result in flat, pavement epithelium.

The walls of *terminal bronchi* (diameter one-one-hundredth to one-two-hundredths of an inch) are composed of a slight amount of connective tissue in which an occasional non-striated muscle-cell and yellow elastic fibre can be distinguished. They are lined with a single layer of flat cells. No definite layers are distinguishable in these bronchi. In a transverse section the lumen would appear circular.

PRACTICAL DEMONSTRATION.

The histology of the bronchi can be studied to best advantage, using tissue from a freshly-killed pig or sheep. Short pieces of tubes, about one-quarter of an inch in diameter, from which most of the lung substance has been cut away, should be hardened quickly in strong alcohol. Transverse sections can be made freehanded, or the tissue may be infiltrated with bayberry tallow or celloidin, and cut with the microtome. Stain with hæma. and eosin, and mount in dammar.

TRANSVERSE SECTION OF PORTION OF BRONCHUS OF PIG. (Fig. 71.)

OBSERVE:

(L.)

1. The **epithelial lining**: (*a*) The **wavy course**. (*b*) Regions occupied by **beaker or goblet cells**. (The letter E in the draw-

ing leads to such a group.) (*c*) The number of **nuclei**, indicating the presence of more than a single layer of cells.

2. The **mucosa**. (*a*) Deeply-stained blue nuclei of the **adenoid tissue** just beneath the epithelium. (*b*) Pink portion of the region below the adenoid tissue. (The longitudinal **elastic fibres** cut transversely.) (*c*) **Blood-vessels**.

3. The **muscular coat**. (*a*) Apparent solution of continuity in places caused by **tubes of mucous glands**. (*b*) The **absence of large vessels in this coat**.

4. The **external layer**. (*a*) Its extent. (It includes the

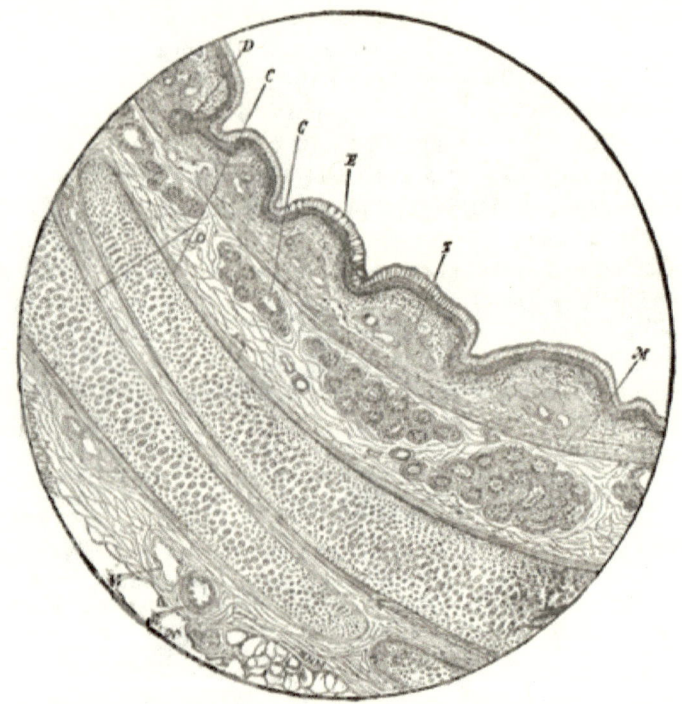

Fig. 71.—Transverse Section of part of the Wall of a Large Bronchus. Lung of Pig. Stained with Hæma. and Eosin. × 60.

E. Epithelial lining. The line from the letter leads to a part of the lining containing large mucous cells.
I. The internal fibrous coat.
M. Muscular coat.
C. Cartilage plates of external fibrous coat.
A. Bronchial artery. The pulmonary artery is not included.
V. Bronchial vein.
N. Nerve trunk.
G. Mucous glands.
D. Obliquely sectioned duct.

remainder of the section.) (*b*) **Large cartilage plates**, C, stained blue. (*c*) **Cartilage cells.** (Note their differing forms and disposition in rows next the surfaces of the plates.) (*d*) **Periosteum**, stained pink. (*e*) **Mucous gland coils.** (They are usually between the cartilage and the muscular coat.) (*f*) Section of **bronchial arteries and veins.** (*g*) Collections of **adipose tissue** on the outer surface. (*h*) Portion or whole of **pulmonary artery** and **medullated nerve trunks** outside of and accompanying the bronchus. (They do not appear in the illustration.)

(**H.**)

5. **Epithelial lining.** (*a*) **Cilia** of columnar cells. (*b*) The **ovoid cells** between the tapering columnar cells. (*c*) The **round cells**, "basement membrane," upon which the columnar cells rest. (*d*) The **goblet or beaker cells**.

6. The **mucosa.** (*a*) The **reticulum of the adenoid tissue.** (Will appear only where the lymph-corpuscles have been accidentally brushed out.) (*b*) The **transversely divided ends of the elastic fibres.** (They appear as a pink mosaic.) (*c*) **Capillaries.** (They may frequently be traced for a considerable distance in their tortuous course.)

7. The cartilage plates. (*a*) **Several cells in a single cavity.** (*b*) The **intracellular network.**

8. The mucous glands. (*a*) That some of the **cells are stained** precisely like the (other) **mucous cells**, the beakers. (*b*) If possible, a gland tube leading up to the lumen of the bronchus. (An ampulliform dilatation is shown in the upper part of the drawing.)

THE PULMONARY BLOOD-VESSELS.

The prominent accompaniments of the bronchus, at the root of the lung, are the pulmonary artery (carrying venous blood) and the pulmonary veins.

The pulmonary artery enters the lung with the bronchus, following in its ramifications, to end in capillary plexuses in the wall of the sac-like dilatations, which are in connection with the ultimate bronchi. The blood is then collected in venules, which unite to form the pulmonary veins. The latter pursue an independent course in their exit, not accompanying the bronchi until the root of the lung (nearly) has been reached.

The bronchial artery (nutrient) enters with the bronchus, supplying its walls and the connective-tissue framework of the lung.

A considerable amount of connective tissue accompanies and supports the organs which enter the lung, and is eventually in connection with the fibrous framework of the organ.

The lung will, therefore, be seen to differ from organs generally, in that it contains *two distinct vascular supplies*, viz.: 1. The *pulmonary* (of venous blood), entering for the purpose of its own oxygenation; 2. The *bronchial* (arterial), which corresponds to the usual nutrient blood-supply of organs.

THE PLEURA.

The lung is completely enveloped with a membrane composed externally of pavement epithelium, while the visceral portion is made up of interlacing fibrous and elastic tissue. The deep or visceral layer of the pleura sends prolongations in the form of septa into the substance of the lung, dividing it into rounded polyhedral compartments or lobules. The interlobular septa have usually become prominent in the human adult from deposits of inhaled carbon in their lymph-channels.

THE PULMONARY ALVEOLI.

The lung is constantly employed in maintaining the integrity of the blood. This is accomplished by the exposure of the latter to a continual supply of atmospheric air. The air is introduced into little *sacs* (termed air-*vesicles* or *alveoli*), in the walls of which the blood is distributed in a *capillary plexus*. The air does not reach the capillaries themselves, inasmuch as they are covered with a layer of flat cells. These cells, constituting the *parenchyma* of the lung, have the power, on the one hand, of selecting such material from the air as may be required, passing it on to the blood in the capillaries; and, on the other, of removing effete materials from the blood, transferring it to the atmospheric contents of the air-sacs for exhalation.

The air-sacs or alveoli are not unlike minute bladders. Their diameter about equals that of a terminal bronchus, viz., from one-one-hundredth to one-two-hundredth of an inch. A group of these alveoli are associated in the manner shown in Fig. 72, their contiguous walls fusing and all opening into a common cavity, the *infundibulum*. The whole is in connection with a terminal bronchus (*vide* Fig. 73). A *primary lobule* having been thus constructed, several are associated and united to a slightly larger bronchial twig, and there results one of the polyhedral lobules, previously mentioned as visible, especially on the surface of the lung. By a repetition of such elements the lung is constructed.

The wall of a pulmonary alveolus or air-sac is composed of connective tissue, supporting the capillary network, with a considerable amount of elastic tissue and an occasional muscular fibre. The whole, as we have said, is lined with a single layer of flat pavement epithelium. The capillary plexus, when filled with blood,

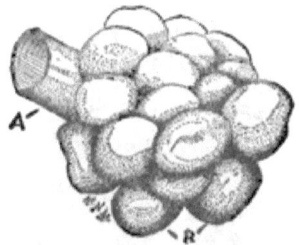

FIG. 72.—DIAGRAM OF AN ULTIMATE PULMONARY LOBULE.
A. A terminal bronchus.
B. The air-sacs or alveoli.

affords the most prominent feature of the wall; but when the vessels have been emptied of their contents, they become very insignificant under the microscope, and the fibro-elastic tissue becomes more apparent. You will have observed that, aside from the vascular supply, the histology of an alveolar wall resembles very closely that of a terminal bronchus, and when the vessels are all empty it is frequently difficult to differentiate them in the mounted section.

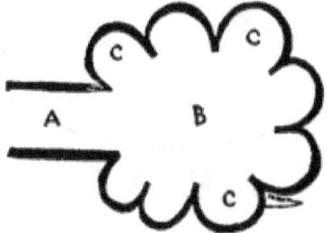

FIG. 73.—DIAGRAM SHOWING AN ULTIMATE PULMONARY LOBULE IN LONGITUDINAL SECTION, SHOWING THE MANNER IN WHICH THE ALVEOLI ARE ASSOCIATED IN CONNECTION WITH A TERMINAL BRONCHUS.
A. Terminal bronchus, entering
B. The infundibulum.
C, C, C. Alveoli.

Fig. 74 shows a single alveolus, the vessels of which have been injected with a solution of colored gelatin. The alveolus has been divided through the middle, and shows as a cup-shaped cavity.

The fibrous marginal walls are indicated with their tortuous capillaries. The epithelial cells lining the bottom are obscured by the opaque capillaries, and shown only between the loops. It is probable that these cells cover the plexus completely as they line the alveoli.

We now encounter an obstacle which will frequently be met in our study of organs. It consists of the difficulty in recognizing in sections the *plan of structure* which we have learned is peculiar to the organ under consideration. For example: A lung has been compared to a tree. The bronchi are the representatives of the branches, and the air-sacs of the fruit. Well, we make a section

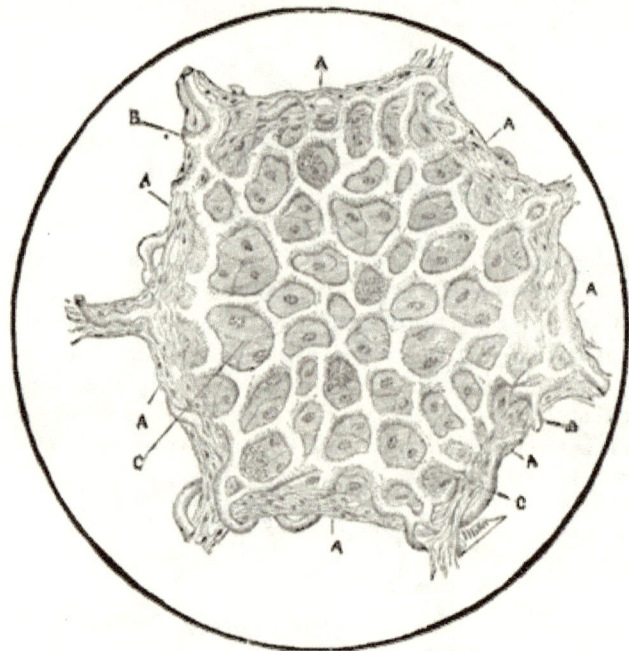

Fig. 74.—Transverse Section of a Single Pulmonary Alveolus. Capillaries injected. Stained with Hæma. and Eosin. × 400.

A, A, A. Walls of the alveolus
B, B. Injected capillaries.
C, C. Pavement cells lining the alveolus. These cells cover the capillaries, but do not so appear in the drawing, as the latter are filled with an opaque injection. The observer is supposed to be above the sectioned alveolus, viewing the cup-shaped cavity.

from human lung—it matters little as to the direction—with every possible care, and the image in the field of the microscope resembles a fragment of ragged lace more nearly than anything else! The arrangement of the tubes and alveoli of the lung has been determined by filling the cavities with melted wax which, when cold, and the tissue destroyed by acid, gives a perfect mould of the organ. A *section* gives us but a single plane, and this fact *must be always borne in mind.*

PRACTICAL DEMONSTRATION.

With a very sharp razor, cut half-inch cubes from pig's lung. Select portions free from large bronchi, with the pleura on one side at least, and harden with strong alcohol. Human lung, as fresh as possible, may be treated in the same manner. The epithelium of the alveoli shows best in young lung. Pieces of fœtal lung are easily hardened, and should be studied with reference to medico-legal work. Lung must be made very hard, or thin sections cannot be cut. If the ordinary 95% alcohol does not harden sufficiently, the process may be completed by transferring the tissue for twenty-four hours to absolute alcohol. The celloidin infiltrating process is well adapted to this structure.

Stain human lung sections with borax-carmine, and pig's with hæma. and eosin. Mount in dammar.

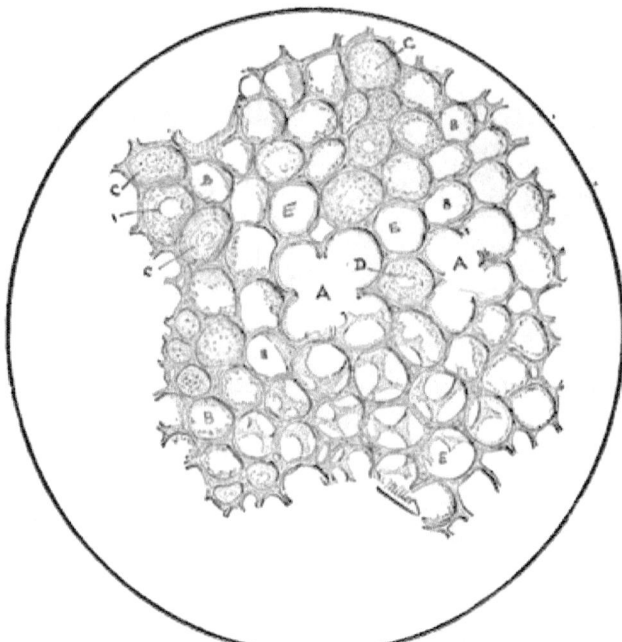

Fig. 75.—Section of Lung of Pig. Stained with Hæma. and Eosin. × 60.
A, A. Infundibula in T. S.
B, B, B. Alveoli; so sectioned as to show the outline only.
C, C, C. Alveoli; so sectioned as to present cup-shaped cavities.
D, D, D. Alveoli; sectioned so as to divide the top (or bottom).
E, E. Terminal bronchi in T. S.

SECTION OF LUNG OF PIG. (*Vide* Fig. 75.)

OBSERVE:

(L.)

1. The large scalloped openings A A, **transversely divided infundibula.**

2. The divided alveoli B B, so **sectioned as to** cut off both bottom and top, and **show no epithelial lining excepting at inner edge of periphery.**

3. The alveoli C C, divided so as to show a **cup-shaped** bottom or top. (The minute granules are the nuclei of the lining cells.)

4. The alveoli D D, so cut as to leave most of bottom or top, **showing an opening in the centre** where the sac has been sliced off.

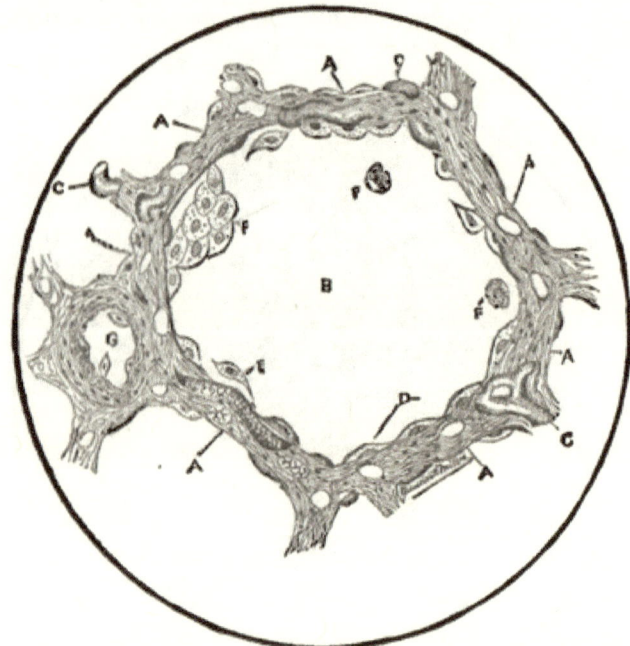

FIG. 76.—TRANSVERSE SECTION OF A SINGLE PULMONARY ALVEOLUS. Stained with Hæma.
× 400.

    A, A, A. Walls of alveolus.
    B. Lumen.
    C, C, C. Capillaries variously sectioned in their tortuous course.
    D. Pavement epithelia intact.
    E. Detached pavement cell.
    F. Detached cluster of pavement cells.
    F'. Granular lining cells.
    G. Pulmonary artery.

5. Openings, E E, which are about the same size and bear a general resemblance to those of Obs. 2. (Note that their internal edges are smooth and not ragged. They are **terminal bronchi**. No larger bronchi have been included in the section.)

HUMAN LUNG, SECTION SHOWING A SINGLE ALVEOLUS. (Fig. 76.)

OBSERVE:

(L.)

1. The **outline of alveolus**. (The alveoli in human lung will show much distortion, as the tissue cannot be secured in perfect condition.)

(H.)

2. The fibrous **wall** A A.

3. The **lumen** B. (The bottom or top has been cut off in making the section.)

4. The **tortuous capillaries** C C, **in the fibrous wall**.

5. The **lining epithelial cells**. (*a*) Those **remaining attached** to the edges of the wall D. (*b*) **Detached** cells E. (*c*) Groups **partly detached** F.

6. The divided **pulmonary artery** G. (A medium-sized bronchus existed in the section immediately to the left of the artery.)

7. Portions of the capillary **plexuses in other alveoli** (not shown in the figure), and especially demonstrable when they may happen to contain blood-corpuscles.

THE LIVER.

This great gland is covered with a fibrous membrane—the *capsule of Glisson*. The capsule is covered with a single layer of irregularly shaped, flat epithelial cells.

Prolongations from the fibrous, visceral portion of Glisson's capsule penetrate the organ from every side, and divide the entire structure into compartments, the *lobules*.

The hepatic lobules are irregularly polygonal in transverse section, and somewhat ovoid vertically. They are about one-twelfth inch in diameter.

Let us first examine the general plan of the vascular arrangement, and later, the minute structure of the lobular parenchyma.

The hepatic blood-supply comes from two sources: 1st, The venous drainage from the chylopoietic viscera collected in the portal vein. 2d, Arterial supply, provided directly from the aorta by the hepatic artery. The portal venous blood is filtered through the liver instead of passing directly to the ordinary destination of such blood (the cava), in order to contribute certain factors to the processes of digestion and metabolism, while the smaller arterial supply is distinctly nutritive. The *hepatic duct* is the common excretory conduit of the bile after its formation by the parenchyma from, mainly, the portal blood.

The scheme of the organ will be understood by reference to Fig. 77, which is purely diagrammatic.

The portal vein enters the liver at the transverse fissure. It divides, subdivides, and, reaching every part of the various lobes, the terminal twigs are seen in the connective tissue of the walls of the lobules.

Branches from these portal termini or *interlobular veins* penetrate the lobular areas, and immediately break up into capillaries, which form an intricate plexus throughout the lobule. The blood from these capillaries is finally collected in a *central or intralobular vein* by means of which it is immediately drained from the lobule.

The central veins, from a number (varying) of the lobules, unite outside of the latter, forming the beginning of the hepatic or so-called *sublobular veins;* and, like vessels from other lobular areas, unite, forming several (six or seven) large *hepatic veins* which, passing in the connective-tissue framework, finally drain the blood from the organ and pour it into the ascending cava as it lies posteriorly in its fissure.

The *hepatic artery* also penetrates the transverse fissure. It accompanies the portal vein in its ramifications, giving off *nutrient twigs to the connective-tissue framework* and to the *walls of the res-*

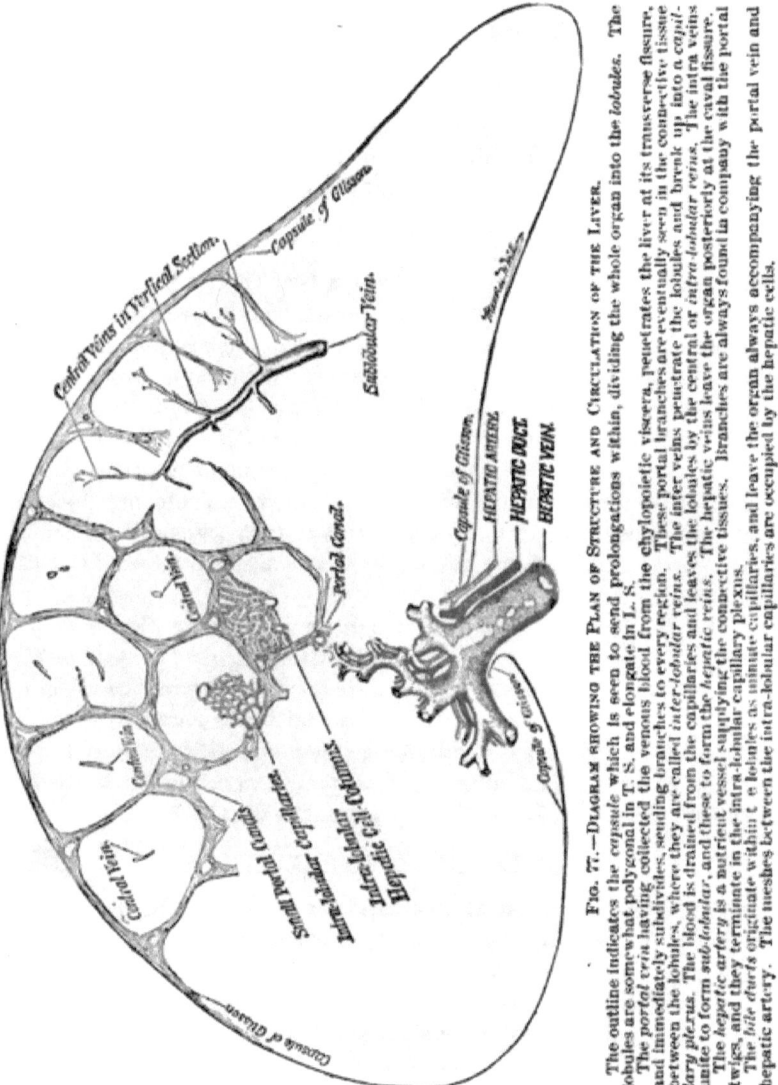

Fig. 77.—Diagram showing the Plan of Structure and Circulation of the Liver.

The outline indicates the capsule which is seen to send prolongations within, dividing the whole organ into the *lobules*. The lobules are somewhat polygonal in T. S. and elongate in L. S.

The *portal vein* having collected the venous blood from the chylopoietic viscera, penetrates the liver at its transverse fissure, and immediately subdivides, sending branches to every region. These portal branches are eventually seen in the connective tissue between the lobules, where they are called *inter-lobular* veins. The inter veins penetrate the lobules and break up into a *capillary plexus*. The blood is drained from the capillaries and leaves the lobules by the central or *intra-lobular* veins. The intra veins unite to form *sub-lobular*, and these to form the *hepatic* veins. The hepatic veins leave the organ posteriorly at the caval fissure.

The *hepatic artery* is a nutrient vessel supplying the connective tissues. Branches are always found in company with the portal twigs, and they terminate in the intra-lobular capillary plexus.

The *bile ducts* originate within the lobules as minute capillaries, and leave the organ always accompanying the portal vein and hepatic artery. The meshes between the intra-lobular capillaries are occupied by the hepatic cells.

sels. The terminal branches, very minute, pour any remaining blood into the venous plexus at the margin of the lobules, thus providing arterial blood for the lobular parenchyma.

The *hepatic duct* is also seen emerging from the transverse fissure. (For sake of clearness, we will trace it from without inward.) It follows the courses of the portal vein with the hepatic artery. Wherever in a section of the organ the portal is divided, the artery and duct will also appear. Bound together with connective tissue, the trio reach the walls of the lobules. The ducts now penetrate the lobules and break up into an exceedingly minute plexus—the *bile capillaries*. This plexus properly *begins in the lobules* and drains the bile as formed, passing it into the ducts in the opposite direction of the portal blood current.

THE PORTAL CANALS.

If it were possible to grasp the vessels as they are found emerging at the transverse fissure, the portal vein, hepatic artery, and hepatic duct, and to forcibly tear them, with their supporting connective tissue, out of the liver, a series of channels or canals would thereby be formed. *A portal canal, then, is the space in the liver occupied by the portal vein, the hepatic artery, the hepatic duct, and the contiguous connective tissue.* Frequently more than one specimen of each vessel is to be seen in the canals. There may be two or three veins, and as many arteries and ducts, associated in a single portal canal. Lymphatic chinks are also abundant in this connective tissue.

From what has been said, it will be understood that *a vessel found by itself* in this organ *must be either a central or an hepatic vein;* and these are easily distinguished, as the former are within, while the *latter are without the lobules* and in the connective-tissue framework. On the other hand, a *group of vessels will indicate a portal canal*, with its large and thin-walled vein, the small thick-walled artery, and, intermediate in size, the duct.

THE LOBULAR PARENCHYMA.

The lobules consist of two capillary plexuses, one containing blood and the other bile. In the meshes of this network, the hepatic cells are located.

The blood capillaries, although extremely tortuous, have a general direction of convergence toward the central veins. This is best seen when the lobules have been divided in a vertical direction.

The bile capillaries are among the smallest canals found in vascular tissues, having a diameter of one-twelve-thousandth of an inch. They pursue a direction in the human liver, as a rule, at *right angles to the course of the blood capillaries*, and are not demonstrable, except with considerable amplification, say × 400, and then only in the thinnest portion of the sections. They are, properly speaking, merely minute channels in the parenchyma, and have, it is believed, no wall.

The hepatic cells are polyhedral, about twice the size of a white blood-corpuscle, say one-one-thousandth of an inch, usually with a single nucleus and with granular protoplasm, frequently containing minute fat droplets and granules of yellow pigment. The existence of a definite limiting membrane has been questioned, as far as the cell of human liver is concerned, although such structure can be shown in many of the lower animals.

The physiological plan of the intralobular structure is expressed in the diagram, Fig. 78. The blood is brought into relation with

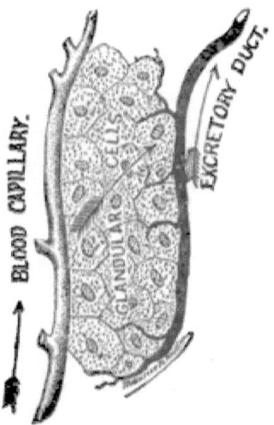

Fig. 78.—Diagram illustrating the Intra-lobular Histology of the Liver.
The hepatic cells are connected in columns between the blood-capillaries. The cells are endowed with the power of selecting, especially, such materials from the blood as are necessary for the manufacture of bile. Having accomplished this, the secreted fluid is given up to the bile-capillaries, and by them poured into the ducts, and led out of the liver for subsequent use. The direction of the pressure is indicated by the arrows. This is the histology of gland structures generally.

the lobular parenchyma—the hepatic cells—by the capillary plexus, and the elements necessary to constitute the bile are selected and carried on, to be drained away by the bile capillaries and ducts.

PRACTICAL DEMONSTRATION.

It is best to begin with the liver from a pig. The amount of connective tissue in the normal human liver is very small, and is mainly confined to the support of the interlobular vessels; the boundaries of the lobules are, therefore, poorly defined, and without the previous observation of some well-outlined specimen, I find the student frequently gets but an imperfect notion of the plan of the human organ.

Pieces of liver, say one-half inch square by a quarter of an inch thick, are hardened by twenty-four hours' immersion in strong alcohol. Larger pieces may be prepared with Müller's fluid. Sections should be cut with a microtome, care being taken to include the transverse division of some of the medium-sized portal canals. The portal vein, with its accompanying vessels, may be easily distinguished from the solitary and less frequent branches of the hepatic veins. The elements of these canals, and especially the larger ones, are best kept intact by infiltration of the tissue with celloidin; but very fine sections may, with care, be made from the alcohol-hardened tissue. Even free-hand cuts, after some degree of skill has been obtained by practice, will answer very satisfactorily. Stain with hæma. and eosin.

SECTION OF LIVER OF PIG. CUT VERTICALLY TO AND INCLUDING THE CAPSULE OF GLISSON.

(Fig. 79.)

OBSERVE:

(L.)

1. The **capsule of Glisson** C. (Note the prolongations sent into the organ, which divide the entire structure into irregularly polygonal, if divided transversely; and elongated, vertically sectioned areas—the hepatic lobules.)

2. The **central** (intra) **veins** C V. (Note that the figure formed by the division of the vein varies according to the direction of the cut, a circle, oval, or elongated slit, as the lobules have been sectioned transversely, obliquely, or vertically.)

3. The **hepatic veins** H V. (Those shown in the section are undoubtedly sublobular. It must be remembered that *sub* applied to these vessels is misleading, as the lobules are situated on every side, as well as above the sublobular veins.)

4. The **portal canals** P C. (Even the smaller ones, I, are readily differentiated from areas containing hepatic veins, inasmuch as a *group* of vessels can be distinguished—the hepatic veins running *solus*.)

5. The **portal veins** V. (Observe that they usually present as the largest element of the canals. Note their thin walls, the same fusing insensibly with the surrounding connective tissue. They not infrequently contain blood-clots, with deeply stained scattering white corpuscles, appearing with this amplification as dots or granules.

6. **Hepatic arteries** A. (The larger examples may be deter-

mined by their thick muscular media and the wavy pink line—the fenestrated membrane. Several may be seen in a single canal.)

7. **Hepatic ducts** D. (These are lined with cylindrical cells, hexagonal in transverse section, and the bold deeply-stained nuclei give the ducts marked prominence even with the low power. Indeed, the smaller portal canals are frequently differentiated by this element alone—this being especially true when the structures have been disturbed, and perhaps torn, in the process of mounting.)

8. The **lobular parenchyma**. (The arrangement of the hepatic cells, forming branching columns, is merely indicated—with

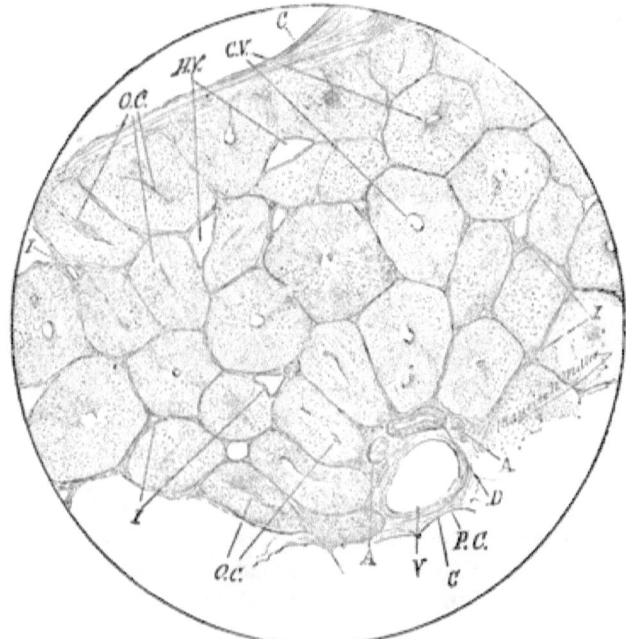

FIG. 79.—LIVER OF THE PIG SECTIONED AT RIGHT ANGLES TO GLISSON'S CAPSULE. Stained with Hæma. and Eosin. × 60.

    C. Capsule of Glisson.
    C. V. Central veins.
    O. C. Oblique section of central veins.
    I, I, I. Inter-lobular veins. (In small portal canals.)
    P. C. A large portal canal.
    A. A. Hepatic arteries.
    D. Hepatic duct.
    V. A portal vein.
    C. Connective tissue from Glisson's capsule.
    H. V. Hepatic veins—probably sub-lobular.

the low power—by their deeply-stained nuclei presenting granular areas within the lobular boundaries. Still, by careful attention, the elements will be seen to radiate more or less distinctly from focal points—the central or intra-lobular veins.)

(**H.**)

9. The **portal veins.** (Note the fusing of the wall with the surrounding tissue—it being extremely difficult to find the line of demarcation.)

10. The **lymph spaces** in the connective tissue of the portal canals. (Note, in those which are better defined, the nuclei of the endothelium. Do not confound these lymphatics with small veins as the latter present a tolerably defined wall, while the lymphatic chinks appear like rifts in the connective tissue; it would be difficult to make this distinction without the endothelial cells.)

11. **Hepatic arteries.** (On account of its solidity, the liver will enable the student to secure sections of blood-vessels presenting the typical structure more nearly than the specimens obtained from the organs heretofore examined.) Note (*a*) the **elongate nuclei** of the sarcous elements **of the media;** (*b*) the **fusing of the adventitia** with the connective tissue surrounding the artery; (*c*) the sharply defined **outer boundary of the intima**—the **fenestrated membrane,** which, from the action of the hardening agent, has contracted the elastic fibres and detached (*d*) the **endothelial cells.** (Inasmuch as the lining cells of small arteries are very frequently partly detached in alcohol-hardened tissue, they may simulate columnar cells. A like appearance is often presented when an artery has been sectioned obliquely, by the projecting muscle-cells of the media.)

12. **Hepatic ducts.** Note: (*a*) The **lining cylindrical cells.** (*b*) The **nuclei** of these cells (as a rule, perfectly spherical; and, in transections arranged in a circle, affording an appearance perfectly characteristic). (*c*) **Mucous glands** in the wall of the larger ducts, lined with large nucleated columnar cells, precisely like those lining the duct-lumen; and, hence, liable to be mistaken for small ducts. (The tube carrying the mucus secreted in these pocket-like glands does not pass directly into the lumen of the duct, but runs along obliquely, much like glands in the bronchi. Not infrequently the glands possess no proper efferent tube, but are mere depressions or diverticula in the thick wall of the bile duct.)

13. The **lobular parenchyma.** (Single cells, partly detached, may be found about the edges of the section.) Note: (*a*) The somewhat **polygonal figure;** (*b*) the **nucleus;** (*c*) **nucleoli;** (*d*) **fibrillated,** mesh-like **cell body;** and (*e*) an apparent **cell wall.**

(The arrangement of the lobular parenchyma will be noted in connection with the human liver.)

HUMAN LIVER.
PRACTICAL DEMONSTRATION.

The sections from which the illustrations have been drawn were made from material hardened in Müller's fluid. The tissue was then cut, the sections washed by six hours' maceration in water, after which they were treated successively with alcohol Nos. 3, 2, and 1, stained with hæma. and eosin, and mounted in dammar. This treatment aids greatly in the demonstration of the blood capillaries, as the contained blood-corpuscles, in consequence of some change effected by the chromium salt, take the eosin deeply. The nucleoli of cells are also rendered markedly prominent.

Pieces of tissue, a quarter of an inch square by half an inch thick, may be hardened in alcohol. This method will give very excellent results, providing the sections be cut as soon as the hardening process has become complete. Stain as above.

For the demonstration of the isolated hepatic cells, scrape the cut surface of a piece of hardened liver with a scalpel, and throw the scrapings into a watch-glass of hæma. After a few moments, drain off the stain, and brush the stained tissue elements into a test-tube nearly filled with water. Change the water two or three times; and when clear, add a few drops of eosin solution. Allow the eosin to stain for a moment only; decant, drain, and fill the tube with alcohol. After ten minutes the spirit may be drained off, and the tube partly filled with oil of cloves. A drop of the sediment may then be placed upon the slide, the bulk of the oil removed with paper, and the mounting completed by adding a drop of dammar and the cover glass. I am in the habit of keeping this tissue in the oil, from year to year, for use in my classes. If the oil be pure, and the washing thorough, the staining will remain unaffected for certainly two or three years.

SECTION OF HUMAN LIVER.

Cut at right angles to the surface, and stained with hæma. and eosin.

(Fig. 80.)

OBSERVE:

(**L**.)

1. The **imperfectly outlined lobules** (in consequence of the absence of interlobular connective tissue).

2. The **fusing of the lobules.** (At points like B B, it is impossible to say just where one lobule ceases and the contiguous one begins.)

3. The **central** (or intralobular) **veins** A A—(frequently appearing as mere slits on account of the direction of the cut).

4. The **portal canals** G G. (These are readily detected on account of the deeply-stained nuclei of the cells lining the hepatic ducts.)

(**H.**)

5. **Portal canals** (too small for demonstration of the several elements, but always distinguishable by the bile-duct cells).

6. The **larger portal canals** C. Note: (*a*) The large **thin-walled vein** D; (*b*) The **duct** E; (*c*) The **artery** F.

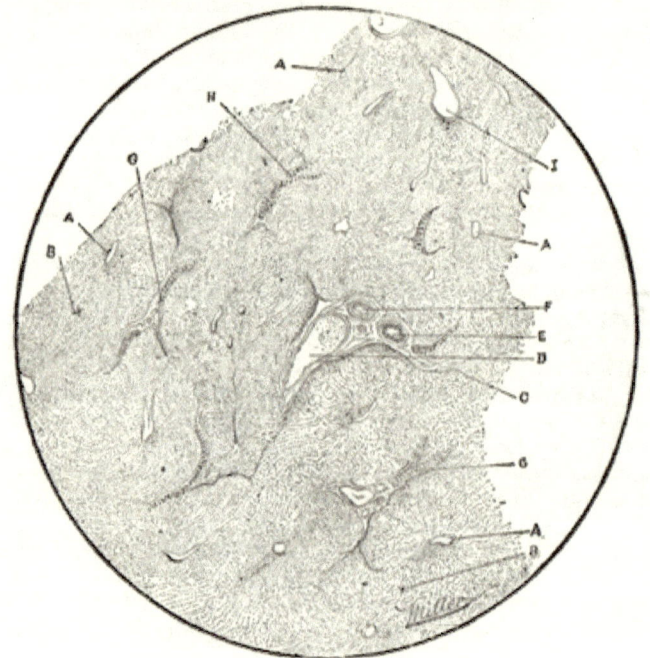

FIG. 80.—SECTION OF HUMAN LIVER.
Stained with Hæma. and Eosin. × 60.

A, A. A. Central veins sectioned generally at right angles to the lobule.
B, B. Points where adjoining lobules coalesce. Illustrating the difficulty of outlining the lobules in normal human liver.
C. Connective tissue of a portal canal.
D. Large interlobular vein.
E. Hepatic duct belonging to C.
F. Hepatic artery of C.
G, G. Smaller portal canals.
H. Small hepatic ducts—always recognizable by the deeply hæma.-stained nuclei of their lining cells.
I, I. Hepatic—sublobular—veins.

7. The tortuous course of the **hepatic cell-columns** as compared with the same in the section previously studied.

8. The **hepatic veins**. (Observe their infrequency compared

with the sections of the portal veins. Note the small amount of connective tissue around them—greater, however, than that about the central veins.)

ELEMENTS OF A PORTAL CANAL. From previous Section.
(Fig. 81.)

OBSERVE:

(H.)

1. The **portal vein** V. (Note the nuclei of the few **endothelia** remaining, and the **corpuscular elements of the blood**

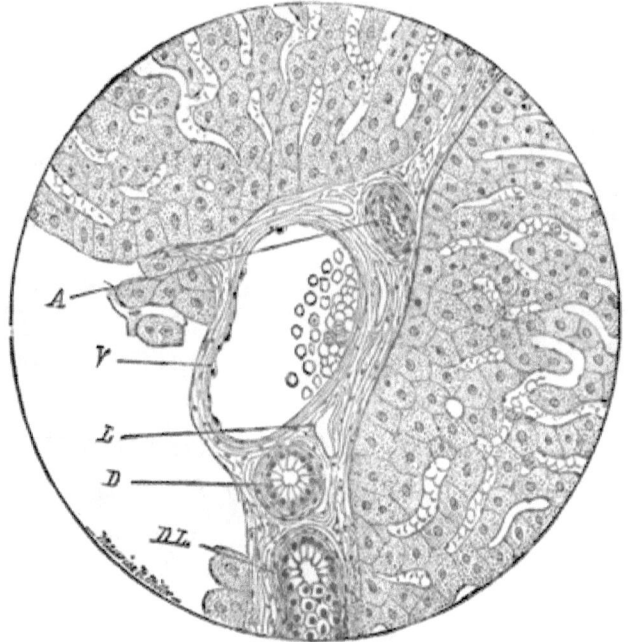

FIG. 81.—SECTION OF HUMAN LIVER SHOWING THE ELEMENTS OF A PORTAL CANAL. Stained with Hæma. and Eosin. × 400.

A. Hepatic artery.
V. Portal vein—interlobular.
D. Hepatic duct in T. S.
D. L. Hepatic duct in L. S.
L. Lymph space.
The lobular parenchyma of contiguous lobules will be seen on the right, and above the canal.

in the lumen of the vein. Observe that the white corpuscles are scanty, and deeply stained, and that many of the colored corpuscles are granular, and show loss of **pigment** from action of the alcohol.)

2. The **hepatic artery** A. (In the human liver, the portal canals frequently carry a number of arteries and ducts, instead of one of each, as shown in the one selected for the illustration. The arteries can nearly always be differentiated by the clear wavy line of the fenestrated membrane. Should the section have been in a longitudinal direction with reference to the vessel, look for the elongate nuclei of the smooth muscle-cells of the media, some running with the artery—the longitudinal—and others at right angles to its course—the circular fibres.)

3. The **hepatic duct** D. (Observe the **thickness of the wall**, depending, of course, upon the diameter of the duct itself—and the presence of **connective tissue** supporting scattering **non-striped muscle-cells**. Note the beautiful, clear, **columnar cell-lining**. That these cells are **polygonal in transverse section** is demonstrable at D L, where the duct has been cut in a longitudinal way, and the cells are seen from above.

4. The **connective-tissue element** of the canal, reaching out in various directions between the adjacent lobules.

5. **Lymph spaces** or chinks L. (Note the stained nuclei of the endothelia.)

6. **Nerve trunks.** (In the larger canals bundles of medullated nerves may be frequently seen. They are not shown in the accompanying illustration.)

THE LOBULAR PARENCHYMA. (Fig. 82.) STAINED CELLS FROM HUMAN AND PIG'S LIVER.

OBSERVE:

(H.)

1. **Isolated hepatic cells** A, A. Note the **large, variably-sized nuclei**, their **nucleoli**, and the **granular protoplasm** of the cell-body.

2. **Groups of cells** forming portions of the hepatic cell-columns as at C.

3. **Cells containing fat globules** D. (This is not necessarily a pathological process, although exactly resembling one, but the physiological storing of hydrocarbons.)

4. **Doubly nucleated cells** B.

THE LOBULAR PARENCHYMA CONTINUED. 117

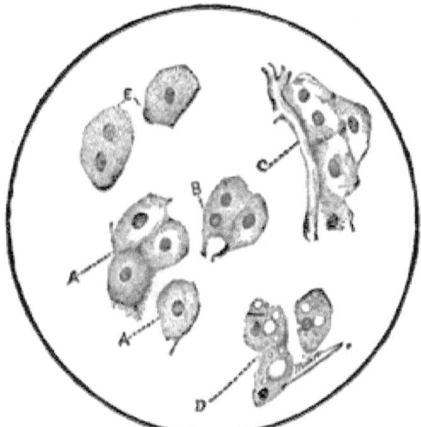

FIG. 82.—ISOLATED HEPATIC CELLS. Stained with Hæma. and Eosin. × 400.
    A, A. Cells from human liver.
    B. Cells from same, showing below a blood capillary in T. S.
    C. A blood capillary with part of a column of cells.
    D. Human liver cells in a condition of fatty infiltration.
    E. Isolated cells from liver of pig, showing intracellular network.

THE LOBULAR PARENCHYMA CONTINUED. SECTION OF HUMAN LIVER.

Fig. 83. (Having found with (L.) a typical lobule in transverse section,)

OBSERVE:

(H.)

1. The **central vein** C. V. (Note the exceedingly delicate wall and search for a trunk of the intralobular plexus in its connection with this vein.)

2. The **blood capillaries in longitudinal section, B, C.** (Observe their exaggerated **tortuosity, bifurcation,** and **anastomoses.**)

3. **Blood capillaries in transection,** T. S. (Should the capillaries be filled with blood, this demonstration will be greatly aided.)

4. **Hepatic cell columns,** H. C. (Note the difficulty with which these can be traced for any great distance, on account of their **irregular and twisted course** throughout the lobule. Observe that the lobules are composed largely of tortuous blood capillaries, between which the hepatic cell-columns are placed. Note the manner in which the **cells are disposed around the blood capillaries,** as at T. S.)

5. **Bile capillaries, D.** (These are rather difficult of demonstration in the human liver. The section should be extremely thin, and a higher power than we ordinarily use will be required. They

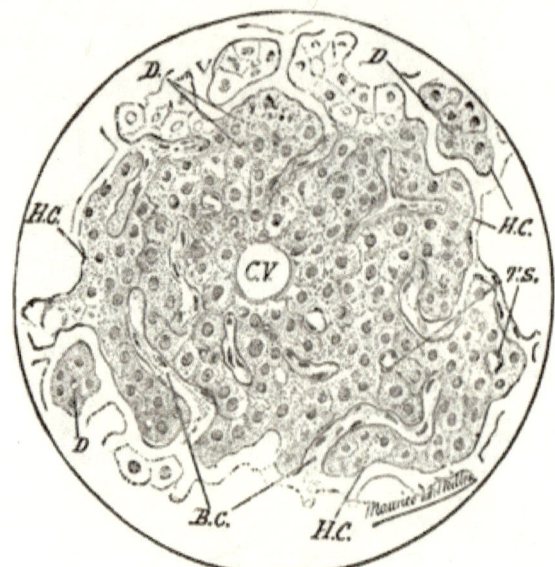

Fig. 83.—A Single Lobule from Human Liver.
Transverse section. Stained with Hæma. and Eosin. × 400.
C. V. Central vein of the lobule.
B. C. Blood capillaries in L. S.
T. S. The same in transverse section.
H. C. Columns of hepatic cells.
D. Bile capillaries.

are best made out at the junction of three or four cells, where the bile capillary has been divided transversely.

THE LOBULAR PARENCHYMA, CONCLUDED. ORIGIN OF THE BILE DUCTS. SAME SECTION AS BEFORE.

(Fig. 84.)

OBSERVE:

(H.)

1. The **connection between the intralobular bile capillaries and the** marginal or intralobular **bile ducts.** (The manner of connection between the above is as follows: The bile capillaries are merely channels between the hepatic cells, and run, as a rule, at right angles to the blood capillaries. They are, I believe,

THE LOBULAR PARENCHYMA CONTINUED. 119

in the human liver, destitute of a wall. As these channels approach the marginal part of the lobule, the hepatic cells surrounding the capillary are seen to change their form. *They elongate, getting thinner, gradually losing their form as hepatic cells, and assume a columnar type. At the same time, a few fibres of connective tissue are thrown outside the modified hepatic cells, and a bile duct results.* The hepatic cells become, insensibly, the columnar

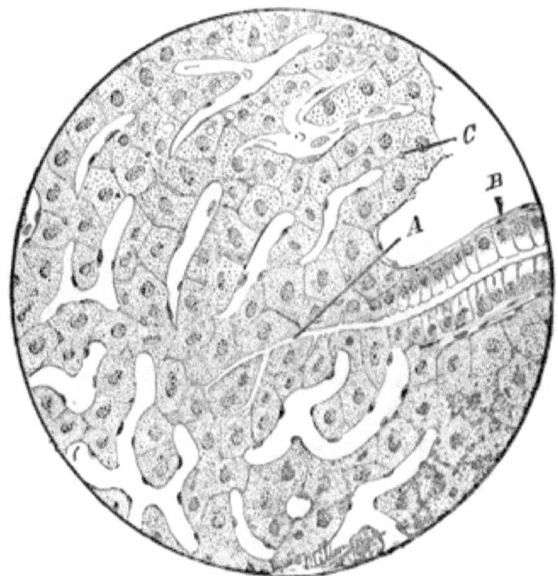

Fig. 84.—Portion of the Periphery of an Hepatic Lobule showing the Origin of a Bile Duct.

Stained with Hæma. and Eosin. × 400.

A. Bile capillaries in longitudinal section.
B. Bile duct. The bile capillaries are simply chinks between the hepatic cells. In order to the formation of a duct, the hepatic cells are altered in shape, elongated, and eventually become the lining cells of the duct. A little connective tissue, thrown around the outside, completes the structure as seen at B.
C. Bile capillary in transverse section. The larger clear spaces are blood capillaries.

cells lining the duct. This is shown in the illustration rather diagrammatically. Its demonstration requires much patient study and search. The duct is best traced backward, as these are readily found.)

THE KIDNEY.

The kidney is as singular in structure as in function. Although developed in lobular form, little trace of this remains in the adult organ.

The kidney consists, essentially, of an intricate system of blood-vessel plexuses, in intimate relation with a system of urine tubes—the whole supported by a small amount of connective tissue.

The accompanying drawing (Fig. 85) will serve to give an idea of the gross plan or scheme of the structure—remembering that the illustration is only a *diagram*.

On making a vertico-lateral section, on the median line, the following appears:

The kidney is invested with a *fibrous capsule*, which is connected with the parenchyma by very delicate prolongations of its connective-tissue fibrillæ. This capsular investment is in connection, above, with the supra-renal bodies; and, on the inner border, with the vessels, etc., which enter and leave the organ at its hilum. The ureter, penetrating the areolar tissue which (containing much fat) presents at the hilum, may, for clearness of description, be traced backward into the kidney. This tube expands into the pelvis, and reduplications of its wall imperfectly divide the pelvic area into three compartments, or *infundibula*.

Each infundibulum is subdivided again, imperfectly, into several pockets or *calyces ;* and into each calyx may be seen, peeping from the kidney substance, a *papillary eminence* or apex of a cone—the *pyramids of Malpighii*. The pelvis is lined with a variety of transitional or imperfectly stratified epithelium, which will be described hereafter.

The blood-vessels, lymphatics, etc., pass in at the hilum, outside the ureter, pelvis, and infundibula. The artery divides into numerous branches which are seen in the diagram passing outward, between the Malpighian pyramids. The renal vein pursues much the same course, the main trunks lying side by side.

On examining a section of the kidney, made in the direction indicated in Fig. 85, a division will be manifest of an outer portion, bounded by the capsule externally, of granular texture, containing the blood-vessels, etc. This is called the *cortex*. Within the cortical portion there appear a number of pyramidal masses—whose apices we have previously seen—of finely striated texture—the medullary or *Malpighian pyramids*. The cortical substance projects itself between the pyramids, completely isolating them, forming the *cortical columns*.

Again observing the outer cortex, it will present narrow, light-colored lines, which converge toward the pelvis; and, eventually, pass into, and become a part of the Malpighian pyramids. These

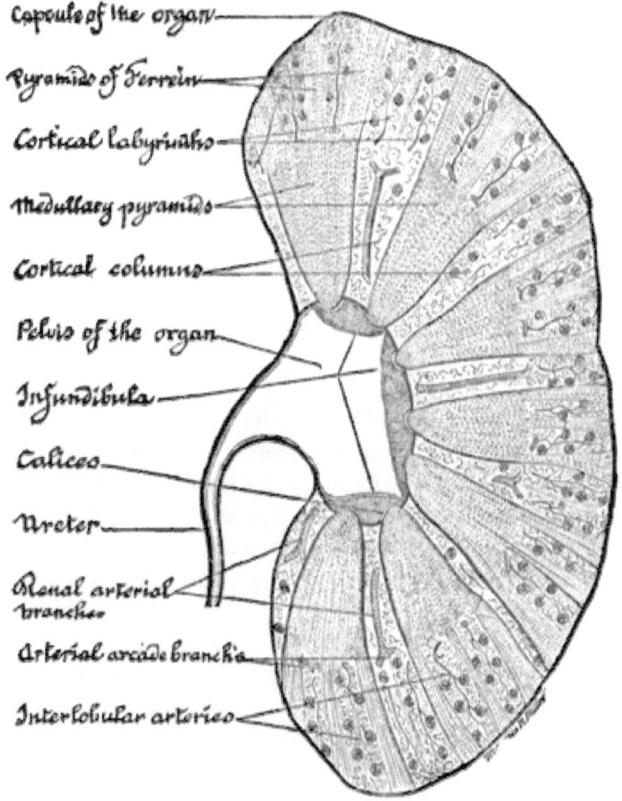

FIG. 85.—DIAGRAM SHOWING THE PLAN OF STRUCTURE OF THE HUMAN KIDNEY.

light areas, made up of urine tubules, are the *pyramids of Ferrein*, or, as sometimes called, the medullary radii.

The darker spaces between the pyramids of Ferrein are called *labyrinths*.

The gross elements, to be understood before we proceed, then, are:

    1. The *capsule* of the kidney.
    2. The *ureter*.
    3. The *pelvis* with its three infundibula. The subdivision of

each infundibulum into several calyces. Each calyx the site of the apex of a Malpighian pyramid.

4. The *blood-vessels* entering and leaving the hilum. Their subdivision outside the pelvic lining, and final passage into the kidney substance in the cortical columns.

5. Division of kidney substance into *cortex* and *medullary* or *Malpighian pyramids*.

6. Penetration of cortical tissue inward between pyramids of Malpighii—constituting the *cortical columns*.

5. The *pyramids of Ferrein*.

6. *The labyrinths.*

In the domestic animals there are no cortical pyramids—the pyramids of Malpighii coalescing, as it were—thus presenting a true medulla.

I have remarked that the kidney is made up largely of urine-carrying vessels (the tubuli uriniferi) and blood-vessels. We will first study the tubular system, reserving for the present the consideration of the blood-vessel arrangement.

THE TUBULI URINIFERI.

The urine-carrying tubules commence in the cortex, and, after taking a very circuitous route with frequently varying diameter, the tubes end at the apex of the pyramids of Malpighii, where they pour their contained urine into the calyces. The urine then overflows into the infundibula, and is finally drained from the pelvis by the ureter.

We shall begin with a single typical tube; and, understanding its histology, we can build up the organ, by simply multiplying this element.

A uriniferous tube, or tubule, commences in the cortex in a labyrinth (between the pyramids of Ferrein), as a thin-walled ($\frac{1}{3000}''$) sac ($\frac{1}{125}''$). This vesicle, with contents, is a *Malpighian body*; and its wall is called the *capsule* of the same, or the capsule of *Bowman*. It is made up of connective tissue and is the thickest part of the uriniferous tube wall or membrana propria, the remaining portion being thin and homogeneous.

From one side of this, the expanded beginning of the tube, a narrow neck ($\frac{1}{1000}''$) is projected, which immediately widens ($\frac{1}{500}''$) into a tube—the *proximal convoluted*. This tube (or this portion of the tube) pursues a very tortuous course, always keeping between Ferrein's pyramids, and finally approaches the base of a Malpighian pyramid. Here it assumes an irregular spiral form—the *spiral tube* ($\frac{1}{500}''$).

The tube suddenly narrows ($\frac{1}{2500}''$), becomes straight, and passes into a pyramid of Malpighii. It reaches sometimes just into the pyramid, more frequently, however, passing deeper than this —often descending two-thirds of the distance to the apex; and is called the *descending limb of Henle*. Henle's descending limb sud-

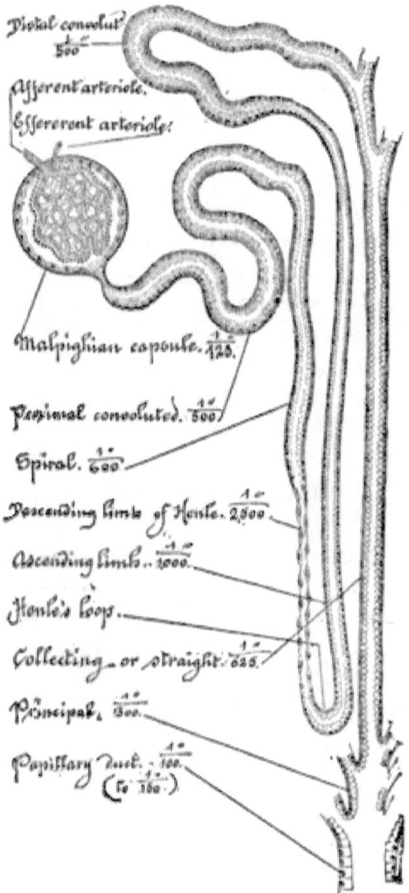

Fig. 86.—Diagram Showing the Divisions of a Kidney Tubule.

denly turns upon itself, forming a *loop;* and, widening ($\frac{1}{1000}''$), returns upon its course as the *ascending limb of Henle*. It again enters the cortex, keeping in a pyramid of Ferrein, and passes outward until it approaches the outer limit of the cortex, near the capsule of the kidney. Here the ascending limb of Henle widens ($\frac{1}{500}''$), forming the *distal convoluted*, which pursues a tortuous course in the outer cortex. The distal convoluted then re-enters a pyramid of Ferrein, narrows ($\frac{1}{875}''$), and passes a second time into a Malpighian pyramid, under the title of *straight* or *collecting* tube, or tube of *Bellini*. The last, after reaching very nearly to the apex of the pyramid, unites with others of a like character, and forms *principal tubes* ($\frac{1}{300}''$). Several principal tubes unite to form a *papillary duct* ($\frac{1}{105}''$). From 100 to 200 of the last open upon the surface of the apical portion of a Malpighian pyramid.

It must be borne in mind that, in describing the tubular system, although such terms as "convoluted tube," "looped tube," etc., are employed, these are not separate tubes, but only *names applied to different portions of one long tube*. A single tubule, then, commences at Bowman's capsule, becomes narrowed like the neck of a flask; courses as the proximal convoluted and spiral; descends into, turns, and emerges from a Malpighian pyramid, as Henle's looped portion; reaches the extreme cortex, and swells as the distal convoluted; and here ends as a single or isolated tubule and enters a straight tube. The straight tubes receive several distal convoluted termini, at the cortical periphery, and pass in small bundles (forming the pyramids of Ferrein) directly onward toward the apex of a Malpighian pyramid; uniting with one another at very acute angles; the resulting trunks uniting until the tube terminates as a papillary duct.

The tubes are lined with epithelia; and these cell elements constitute the *parenchyma of the kidney*. The lining cells are, *as a rule, of the columnar variety*. Two exceptions are presented, one of which appears in the flattened cells lining *Bowman's capsule*, and the other in a like form, in the *descending limb of Henle's loop*. The parenchyma will receive attention in our practical work.

BLOOD-VESSELS.

The vascular arrangement is complex. The most prominent and essential feature is afforded in the existence of two distinct capillary plexuses.

The renal artery, as already described, sends branches into the substance of the kidney. These pass between the Malpighian pyramids, and *in the cortical columns*. These arterial trunks arch over the bases of the pyramids of Malpighii, forming the *arterial arcade*. From these arches small straight branches are sent outward toward the capsule of the kidney, occupying a position midway between the pyramids of Ferrein, in the labyrinths. The last are the *interlobular arteries*. During their course, they send off side *arterioles* which penetrate the capsule of the Malpighian bodies. Each afferent arteriole breaks up into a capillary plexus— the *tuft* or *glomerulus*. The glomerulus does not entirely fill the

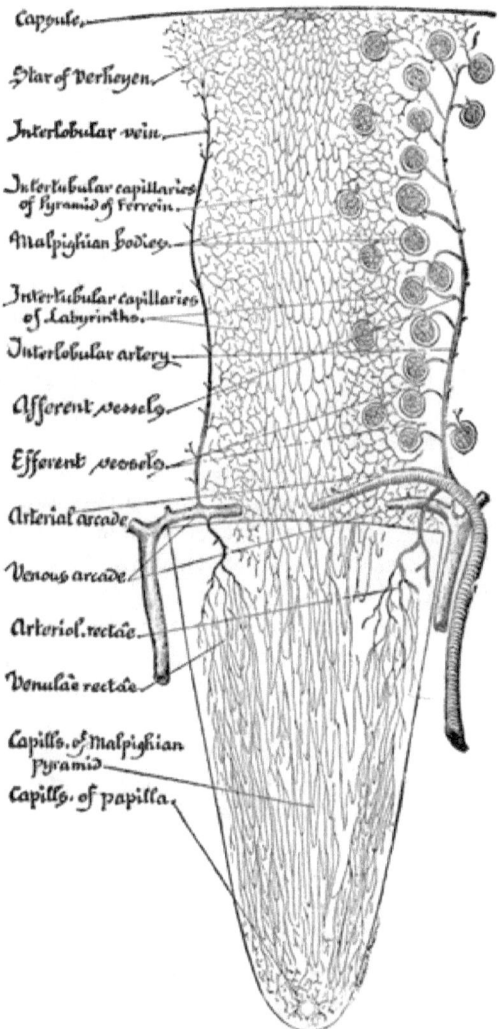

FIG. 87.—DIAGRAM SHOWING THE ARRANGEMENT OF BLOOD-VESSELS IN THE KIDNEY. After Ludwig.

capsule, so that a space remains between the spherical mass of capillaries and the flattened cells lining the body. The glomeruli are enveloped with a single layer of flattened epithelial cells.

The blood escapes from the glomerulus by one or two *efferent arterioles* which emerge from the capsule close to the afferent vessel. The latter is the more noticeable, as it is usually much the largest. The efferent arteriole immediately breaks up into a second capillary plexus, which courses between the uriniferous tubules of the labyrinths and of the pyramids of Ferrein. This second plexus also descends between the elements of the pyramids of Malpighii. From the arteries forming the arcade another set of branches—the *arteriolæ rectæ*—is given off; which, descending into the Malpighian pyramids, provides another and direct arterial supply to the tubular elements by elongate capillary loops.

The course of the venous trunks is not unlike that pursued by the arteries. Interlobular veins pass into a venous arcade; the former lying in the cortical labyrinths parallel with and close to the arteries. In the medulla the venous blood is collected from the capillaries and carried to the bases of the Malpighian pyramids in small veins—venulæ rectæ. The blood from the cortical intertubular capillaries is collected in the interlobular veins.

A peculiar vascular arrangement exists just beneath the capsule of the kidney, consisting of scattered venous plexuses, the *stars of Verheyen*. They contain blood collected from contiguous intertubular capillaries and are in connection with the summits of the interlobular veins.

From what has been said, it will be seen that the cortical and medullary blood-supplies are, to a certain extent, independent of one another. The arteriolæ rectæ provide a vascular supply to the elements of the Malpighian pyramids even after many of the glomeruli have become obliterated by disease.

Nerve and lymphatic elements are not very prominent features in sections of the kidney. Small medullated nerve trunks may be easily demonstrated in transverse sections of the cortex, especially near the bases of the medullary pyramids, where they will be seen, in company with the blood-vessels of the arcades. Lymph channels are also to be seen in the vicinity of the vessels of the hilum, and in the connective tissue of the capsule. The nervous system of the kidney would prove a valuable field of labor, and would well repay the advanced student's patient and earnest investigation.

The histology of the kidney will be better comprehended by a reference to its functioning. The separation from the blood of a quantity of water, together with certain excrementitious matters, is effected, partly in the Malpighian bodies, and partly in the tubules. The vascular tuft—the glomerulus—is covered with a close fitting membrane composed of flat cells. The blood in this plexus parts with a certain amount of its water, which passes through the walls of the capillaries and through the cells covering them. Whether this be due to osmosis or to some selective power of the cells we have no concern—suffice it that certain salts afterward appearing in the urine do not leave the blood at this point. The efferent glomerular arteriole, it will be remembered, breaks into a second capillary plexus, which brings the blood close to the walls of the convoluted tubules. We believe that the cells lining these tubules select from the blood, circulating in the contiguous capillaries, such effete materials as escaped elimination from the glomeruli. Moreover, that some of the water, together with serum albumin, which escaped in the first instance and entered the proximal convoluted tubules, is here returned to the blood by the intervention of the same tubular lining cells which excrete the salts. That in the cells of these tubules there exist currents in opposite direction—one from the intertubular capillaries into the proximal part of the tubule; and one from the dilute urine in the tubule into the capillaries. Without referring to any further work on the part of the kidney, I wish to impress this part of the structural scheme: That the first part of the uriniferous tubule is the prominent excreting part. That the latter portion of the tubule—the portion in the Malpighian pyramids, the straight tubule—is for the collection and drainage of the urine already excreted. *And that between the excreting first part and the draining second part, there exists a narrow looped tubule—the loop of Henle.* The effect of this narrowing and tortuosity of the tubule will be to present a resistance to the outflow of urine from the proximal portion of the tubule. The dilute urine, excreted in the Malpighian bodies, is held back for a while in the proximal convoluted, and time given for the completion and perfection of the excretory processes by the tubular parenchyma.

PRACTICAL DEMONSTRATION.

The human kidney is rarely found in a perfectly normal condition. The demonstration can be made from the kidney of the pig, except as regards certain features. The medullary pyramids do not exist in the domestic animals, and the parenchyma presents very essential differences from the cells of the human kidney. Still, very much can be learned from the organ of the pig, dog, and rabbit. The tissue should be divided so as to permit sections to be made parallel with the medulla, and to include both it and the cortex. The hardening is best by Müller's fluid. Small pieces hardened quickly in strong alcohol, however, stain very finely with hema, and eosin. Very pleasing differentiation may also be secured by staining slowly in weak borax-carmine, clearing with glycerin, and mounting in the same medium.

HUMAN KIDNEY. SECTION PARALLEL WITH MALPIGHIAN PYRAMID. STAINED WITH HÆMA. AND EOSIN.

(Fig. 88.)

OBSERVE:

(Naked eye.)

1. The **thickness of the cortex**, and its **granular appearance** as compared with the medullary portion.

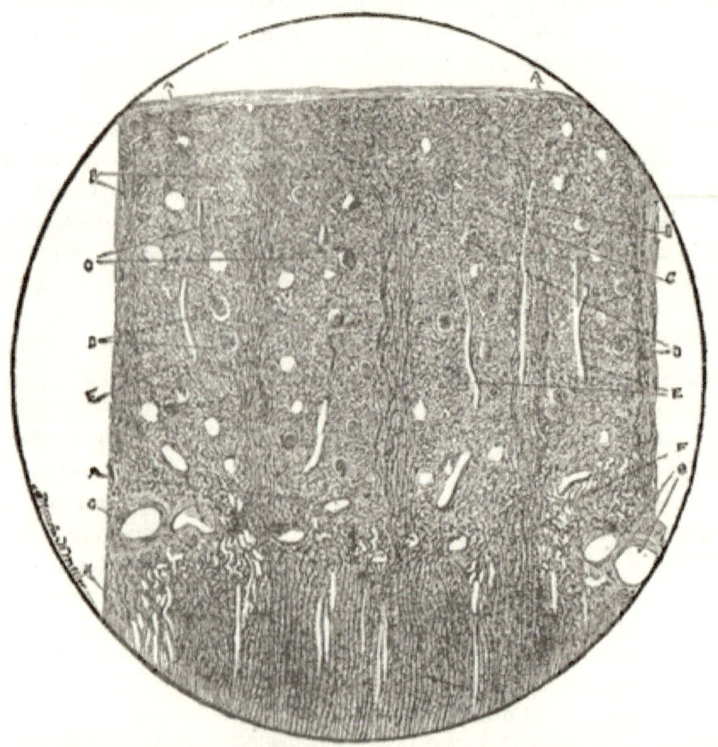

FIG. 88.—SECTION OF HUMAN KIDNEY, CUT PARALLEL TO THE PYRAMIDS OF FERREIN. SHOWING THE CORTEX AND PART OF A MALPIGHIAN PYRAMID. × 30.

A, A. Capsule of kidney.
B, B. Pyramids of Ferrein.
C, C. Cortical labyrinths.
D, D. Malpighian bodies. Many of the glomeruli drop out in the course of preparation, and such empty capsules of Bowman appear as light circular spots.
E, E. Interlobular arteries.   F, F. Boundary region.
G, G. Transverse sections of vessels of the arcades.
H. Base of a Malpighian pyramid.

2. The "**markings of the cortex.**" These consist of alternating light and dark lines, radiating from the bases of the Malpighian pyramids. The lighter masses consist largely of collecting tubes, together with ascending limbs of Henle's looped tubes—otherwise called **medullary radii**. Between these lighter areas the dark **labyrinths** appear; in which, by careful attention, the Malpighian bodies may be made out as minute red dots.

3. A region just outside the medullary pyramids—not as well marked as the outer cortex, in which few Malpighian bodies present—the **boundary region**.

4. The finely striated medullary or **Malpighian pyramids**. (The section will usually include portions of two of the last.)

5. That the **bases of the pyramids** do not appear as a sharply-defined line, but fade into the boundary region; while the union of the latter with the cortex proper is equally ill-defined.

(**L.**) Fig. 88.

1. The **cortical labyrinths**, in which search for—

(*a*) Portions of the **interlobular arteries**, together with the smaller twigs of the **arterial arcade.**

(*b*) The **Malpighian bodies.** (The tuft or glomerulus which, with this power, appears as a granular mass, is wanting in numerous places—as indicated by the empty capsules.)

(*c*) The remaining area occupied largely by the **convoluted tubes**, proximal and distal.

2. The **pyramids of Ferrein.** (Observe that, as they pass into the pyramids of Malpighii, they are well defined, but that they are lost as they approach the region of the capsule of the organ.)

(**H.**) Fig. 89.

1. A **Malpighian body.** (Select, after searching several fields, a specimen which shows either the **afferent** or **efferent vessel** of the glomerulus. It will be very difficult to find a **capsule connected with the neck of a proximal convoluted tube,** as they rarely happen to be so sectioned. You may indeed be obliged to examine a dozen slides before you succeed.) Note—

(*a*) The **capsule** (of Bowman or of Müller). (Observe its thickness, as this becomes important in connection with the pathology of the kidney.)

(*b*) The **flattened cells, lining the capsule.** (Many of them will have become detached in the preparation of the section.)

(*c*) The **glomerulus.** (The great number of nuclei obscures the loops of capillaries. Remember that the nuclei belong partly

130   PRACTICAL MICROSCOPY.

to the vessels, and partly to the flattened cells covering the glomerulus. Endeavor to find **transversely divided loops of the vessels**, showing blood within.)

(*d*) That the **glomerulus does not, entirely, fill the capsule.**
(*e*) That the **tuft is frequently divided.**
(*f*) That the **tuft is usually in contact with the capsule at some one point,** where search may be made for

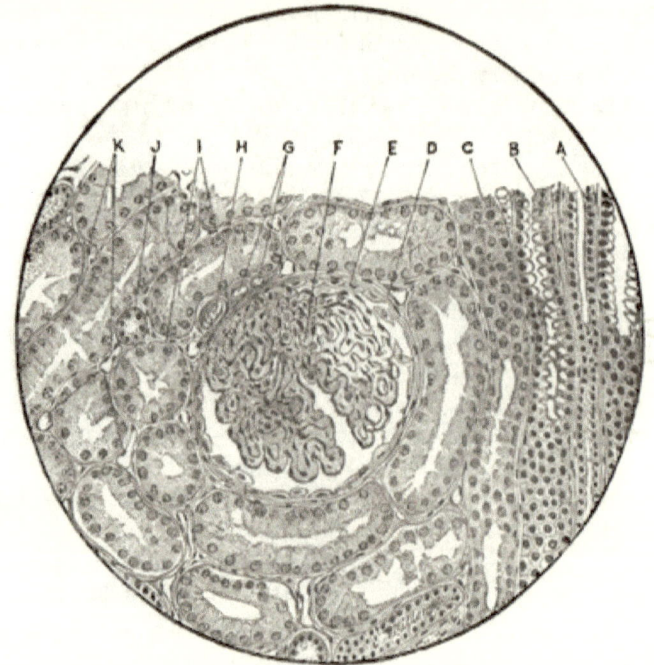

FIG. 89.—PART OF THE CORTEX OF HUMAN KIDNEY. HIGH POWER. SAME SPECIMEN AS FIG. 88. × 400.

A. Ascending limb of Henle's loop.
B. Collecting tubule—longitudinal section.
C. Collecting tubule. The upper part of the tube is not sectioned, but shows the attached bases of the lining cells; and thus simulates pavement epithelium. A, B, and C are in a pyramid of Ferrein.
D. Capsule of a Malpighian body. The emerging tubule is not shown, as the body is in T. S.
E. Flattened lining cells of D.
F. Glomerulus.
G. Efferent arterioles.
H. Afferent arteriole.
I. Convoluted tubules.
J. Ascending limb of Henle.
K. Intertubular capillaries.

(*g*) The **afferent and efferent arterioles.** (The afferent is more frequently demonstrable; and may be differentiated by its large size and the connection, which can often be traced, with the interlobular artery.)

2. **Convoluted tubules.** (The convoluted tubes found just beneath the capsule of the kidney generally belong to the distal variety; and they are not as favorable specimens as the deeper proximal portions, on account of the crowding of the tubular elements in the outer cortical regions. Select a transverse section and observe:)

(*a*) The thin **membrana propria,** or wall of the tube. (It does not appear to be made up of fibrillated connective tissue; but, rather, presents a homogeneous structure. Nuclei, however, may occasionally be seen, which apparently belong to this tissue.)

(*b*) The peculiar **lining cells.** (They are unlike any other parenchymatous elements found in the body. Note that, while they are evidently of the columnar or cylindrical type, they differ greatly in form and size. The protoplasm is hazy, granular, and frequently striated. They take a dirty brick-red hue from the eosin.)

(*c*) **The lumen.** (Compared with the diameter of the tube-wall, the lumen is very small, and presents a stellate figure. The urine, in passing through the tubule, is, consequently, brought in contact with a very considerable portion of the parenchymatous lining.)

3. The **large proportion of the cortical area occupied by the convoluted tubules,** and the exceedingly **small amount of intertubular connective tissue.**

4. The **intertubular capillaries.** (These are exceedingly small, and difficult of demonstration unless they be filled with blood. The nuclei of the endothelial wall are frequently seen. The cells of the convoluted tubules are not infrequently detached from the membrana propria, and the space so formed may be mistaken by the careless observer for longitudinal sections of capillaries. These vessels are much better seen in an injected kidney; although if an organ be selected containing considerable blood, and the corpuscular elements have their color preserved (as in bichromate hardening), the vessels will be easily demonstrated.)

5. **Ascending limb of Henle's loops,** in the cortical labyrinths. (The general course of these tubules is confined to the pyramids of Malpighii and Ferrein; but occasionally one of them may be seen passing in a tortuous course toward the outer cortex, running between the proximal convoluted elements. They are easily recognized by their small size and relatively large lumen. They are lined with short columnar or cuboid cells, which stain deeply blue with the hæma.)

6. The pyramids of Ferrein.

(*a*) **Collecting tubes.** (These will be generally recognized by their large size and the blue color of the staining. They are lined with columnar cells, which are hexagonal in transverse section; and this gives an appearance of pavement epithelium, when they are seen from above or below. Endeavor to find a tube split through the centre longitudinally and note the typical columnar cells, as they project inward from the membrana propria, toward the now open lumen.)

(*b*) **The spiral tubules.** (These resemble somewhat the convoluted tubules, especially as their cells take much the same

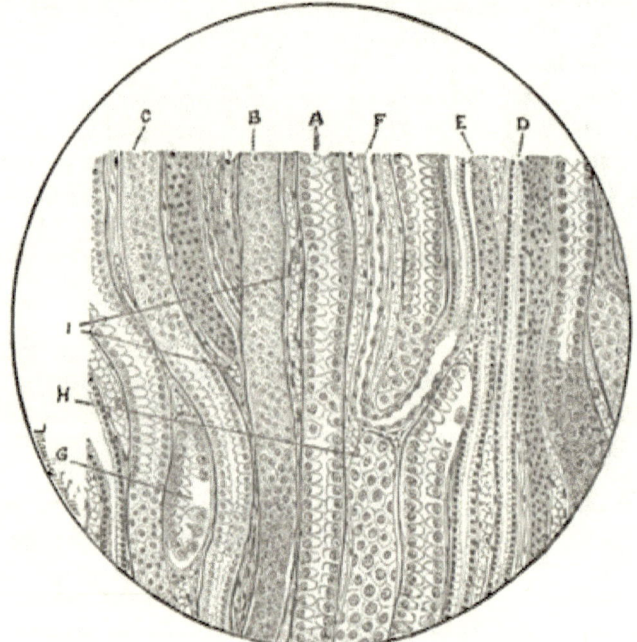

Fig. 90.—Medullary Portion of Specimen Shown in Fig. 88. × 400.
  A. Collecting tubule in L. S.
  B. Collecting tubule from above, showing attached bases of lining cells.
  C. Collecting tubule presenting different appearance of lining cells, according to mode of section.
  D. Ascending limb of Henle's loop.
  E. Same as last. The upper end of the tubule not sectioned.
  F. Descending limb of Henle's loop. Below may be seen the loop and ascending limb.
  G. Oblique section of large collecting tubule.
  H. Basal, attached extremities of cells lining a large collecting tubule.
  I. Intertubular capillaries.

dirty red color. The cells, however, plainly columnar, are large and hexagonal in transverse section. The lumen is small.)

THE KIDNEY. 133

(*c*) **Ascending limbs of Henle's loop.** (These are small tubes and have already been described.)

(*d*) **The intertubular capillaries.** (Inasmuch as, in the specimen under consideration, the vessels of the pyramids are mostly in transverse section, they are not readily made out. Especially is this true if the blood-corpuscles have their color discharged.)

7. **Elements of the medullary portion.** Fig. 90.

(*a*) **Collecting tubes.** (These tubes constitute a large proportion of the medulla of the organ. They have been already de-

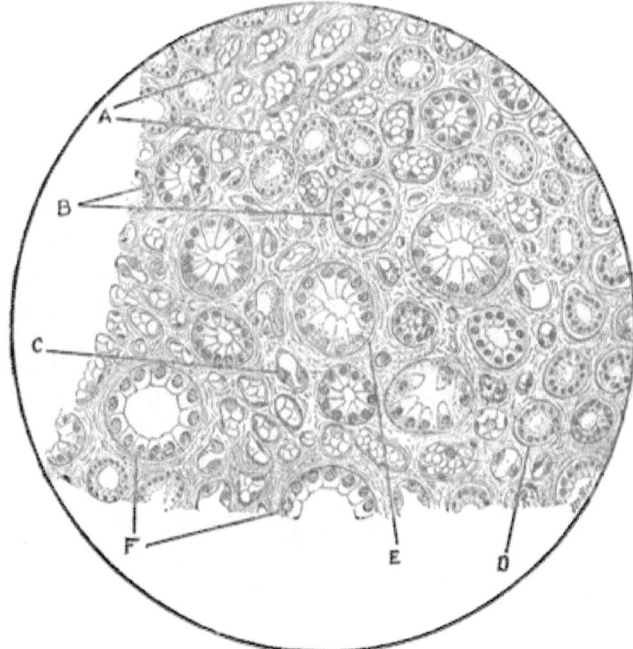

FIG. 91.—TRANSVERSE SECTION OF PYRAMID OF MALPIGHII. SAME TISSUE AS SHOWN IN FIG. 88. Stained with Hæma, and Eosin. × 400.

A. Group of intertubular blood-vessels.
B. Collecting—straight—tubules.
C. Descending limb of Henle's loop.
D. Ascending limb of Henle's loop.
E. Principal—collecting—tubule.
F. Principal tubule. Lower portion near the papillary duct.
The ring of cells will be seen detached from the membrana propria in some instances. This is due to contraction of the tissue during the hardening.

scribed. As the apex of the Malpighian pyramid is approached, and the straight unite to form the principal collecting tubes, these again uniting to form the papillary ducts, the lining cells will be seen to get shorter, and the lumina larger.)

(*b*) **Spiral tubes.** (These can be, in many instances, followed down from the pyramids of Ferrein; and examples are frequently seen very near the pelvis of the kidney in the cortical columns.)

(*c*) **Descending limbs of Henle's loop.** (These tubes are the most difficult of all the tubuli uriniferi to demonstrate. The section must be very thin, and, even then, they may be mistaken for blood-capillaries. Their peculiar feature consists in the wavy lumen, which is produced by the alternate disposition of the lining cells.)

(*d*) **Loop of Henle.** (The loops will be recognized by the curving of the tube. They are lined with short columnar cells which are sharply brought out by the hæma. On account of their course but few complete sections are seen.)

(*e*) **Ascending limbs.** (Conveniently traced from the loop.)

(*f*) **Intertubular blood-vessels.** (Do not mistake tubules containing blood, for capillaries. The human kidney is rarely absolutely normal; and blood is frequently found outside the proper channels. The vessels will be differentiated by the histology of their walls. Quite a number of venules will be seen running in groups in the medulla—the *venulæ rectæ*.

8. **The same elements as in 7** (shown in a transverse section of the middle of a Malpighian pyramid, Fig. 91).

EPITHELIUM OF THE GENITO-URINARY TRACT. URETER, BLADDER, UTERUS, VAGINA, ETC.

The lining membrane of the genito-urinary apparatus is of interest to the medical man; particularly in connection with diseases whose diagnosis may be largely determined by a microscopical examination of urinary deposits.

In the preparation of this subject, I have confined myself, rather closely, to the consideration of such portions of the lining of the genito-urinary tract as may be recognized and differentiated, as they occur in urine and other fluid discharges. The limit prescribed for these pages will permit little beyond this.

It is not always possible to determine the origin of detached cells, for two reasons, viz.: First, certain widely-separated portions are very similarly cell-covered; and, secondly, cells which are the result of proliferation accompanying diseased processes are quite frequently unlike the original type. Still, certain portions of the genito-urinary apparatus have a distinctly characteristic epithelium; and to such will our present notice be directed.

PRACTICAL DEMONSTRATION.

From the body of a (preferably young) human female, as soon as possible after death, remove half-inch cubes of the organs required, observing that the lining is included. The outer portions are of very little moment comparatively. Secure pieces from the os, cervix and fundus of the uterus, the base of the bladder, the wall of the vagina near the cul-de-sac, the ureters, and the pelvis of the kidney.

We desire to prepare the tissue so as to keep the original form of cell elements—to avoid contraction; and the Müller process will accomplish this perfectly. Allow the pieces to remain for two weeks in the bichromate solution, with an occasional change. Complete the hardening in Nos. 3, 2, and 1 alcohol, as usual. Infiltrate with celloidin or bayberry tallow, and let the sections be vertical to the mucous surface. The tissues should not be handled with the fingers, otherwise the epithelial lining cells will be detached. Stain with hæma. and eosin; mount in dammar.

UTERUS AND VAGINA OF THE HUMAN FEMALE AT PUBERTY.

VERTICAL DEXTRO-SINISTRAL SECTION OF THE RIGHT LIP OF THE OS, AND INCLUDING PART OF THE VAGINAL CUL-DE-SAC.

OBSERVE:

(L.)

1. The **outline of the section.** (Commencing at D, Fig. 92, which is placed in the **internal os,** followed downward, out upon

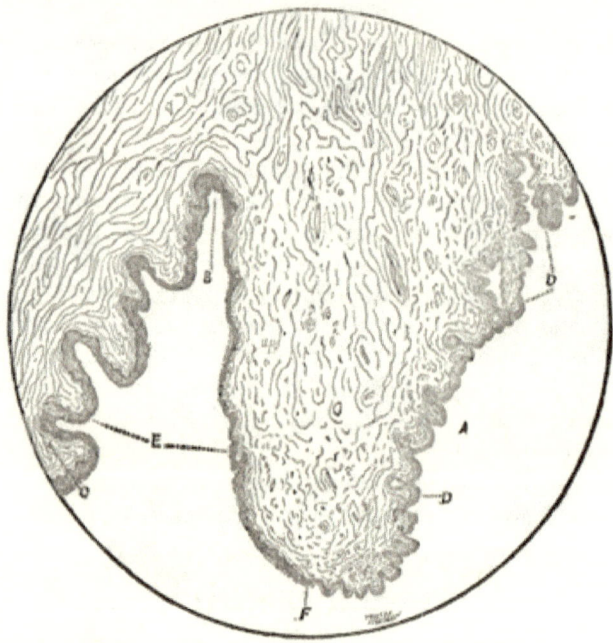

FIG. 92.—VERTICAL DEXTRO-SINISTRAL SECTION OF THE RIGHT-HAND SIDE OF THE OS UTERI. SHOWING THE INTERNAL OS, THE EXTERNAL OS, THE VAGINAL CUL-DE-SAC, AND THE UPPER PORTION OF THE VAGINAL WALL. × 60.

    A   The letter is placed in the internal os.
    B. Vaginal cul-de-sac.
    C. Vaginal wall.
    D. Columnar epithelium of the internal os. In the upper portion the tubular glands are well seen.
    E. Stratified epithelium of the vaginal lining.
    F. Change at the external os from stratified flattened, to columnar epithelium.

the **external os,** curve upward, reaching, at B, the **vaginal cul-de-sac.** Descend along the right **vaginal wall.**)

2. The **irregular surface of the internal uterine wall.** (Caused by longitudinal section of the glandulæ uterinæ or g.

utriculares, branched tubular glands. These are increased in depth during pregnancy, and are most prominent in the lower portion of the organ.)

3. **The epithelium.** (*a*) The **deeply stained layer lining the vagina, cul-de-sac,** and **external os.** (*b*) The **wavy course** of *a* as it covers the irregularly formed and often imperfect **papillæ of the mucosa.** (*c*) The lighter appearance of the **lining of the internal os.** (*d*) **Projection** of the last **into the glands.** (*e*) The **sharp line of separation between** the deeply stained lining common to the **vagina and** the lighter lining of the uterus at the **external os** (Fig. 92, F).

4. The **mucosa of the uterus.** (There are no sharply defined regions in the genito-urinary tract corresponding to the mucosa and submucosa of typical mucous membranes. The arrangement generaly is: 1, an epithelial lining; 2, a subepithelial structure, consisting of a more or less prominent or abundant plexus of capillaries supported by delicate connective tissue, and which corresponds to the **mucosa** of typical structures; 3, loose connective tissue, with more or less muscular tissue, containing larger vessels, not separated from the mucosa by any well-defined line or muscularis mucosæ, which represents the submucosa; 5, the muscular walls proper, consisting of layers in different directions, frequently irregularly disposed and seldom in distinct fasciculæ.)

5. The **mucosa of the vagina** (less distinct than that of the uterus).

6. The **uterine and vaginal walls** (consisting largely of involuntary muscular fibrils, recognized by the elongate and deeply stained nuclei, and containing numerous thick-walled arteries and irregular lymph spaces).

(**II.**)

7. The **uterine epithelium** (Fig. 93). (*a*) That it consists of a **single layer** of cells. (*b*) That the **cells are columnar, not cylindrical.** (*c*) The cells in transverse section are **polygonal.** (*d*) They are **ciliated.** (This demonstration is not always made; but if the section has been properly prepared from uninjured tissue, the cilia will be seen without difficuty, and especially in the depressions where they are somewhat protected.) (*e*) The **cell body** and **nucleus.** (Note the elongate, clear, free portion and the frequent curving of the whole. Near the attached extremities, which often appear pointed, note the small deeply stained nuclei.) (*f*) The **large mucous cells.** (These singular cells appear scattered between the cylinders, with a clear bulging body, often six times the

breadth of the ordinary elements.) (*g*) The **absence of** any special **basement membrane.**

8. The abrupt **transition from columnar to flattened cells** in the epithelium of the external os. (*a*) The **shortening of the columnar cells** as the point of change is approached. (Sections must be examined until one is found showing this point well. The illustration (Fig. 93) is not exaggerated, and a properly cut and selected specimen must exhibit clearly the last columnar and the

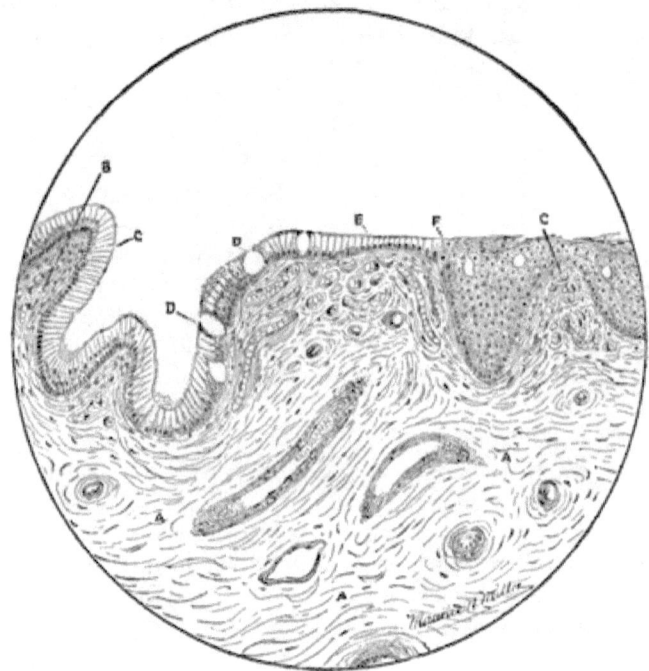

Fig. 93.—External Os of Fig. 92. More Highly Magnified. × 400.
  A. Muscular tissue of the os uteri, with numerous blood-vessels.
  B. Capillary plexuses of sub-epithelial tissue—mucosa.
  C. Ciliated columnar cells covering the os.
  D. Vacuolated cells.
  E. Shortening of the columnar cells prepartory to
  F. Change from typical uterine epithelia—ciliated columnar cells—to flattened, stratified cells.
  G. Papillary structure of the mucosa of the external os, after the change of epithelium.

adjoining flattened cell. I know of no location in the human body where the change in form of cell covering approaches this in abruptness.)

9. The **vaginal epithelium** (Figs. 93 and 94). (*a*) That it is of the **stratified** variety. (*b*) The **deepest line of cells** follow-

ing the sinuous line formed by sectioning the papillary mucosa. (c) That the cells are more or less **flattened**. (d) That their edges, excepting those of the surface, are **serrated**. (The union is by a cement between the interdigitating cell bodies.) (e) The **change** in form **as the surface is approached**. (f) The **surface cells**. (These are very much flattened, and so fused as to resemble, in longitudinal section, fibres.) (g) **Detached surface cells**. (At H, Fig. 94, these are shown in plan, having been torn off; those intact are, of course, seen in profile. Fig. 97 represents

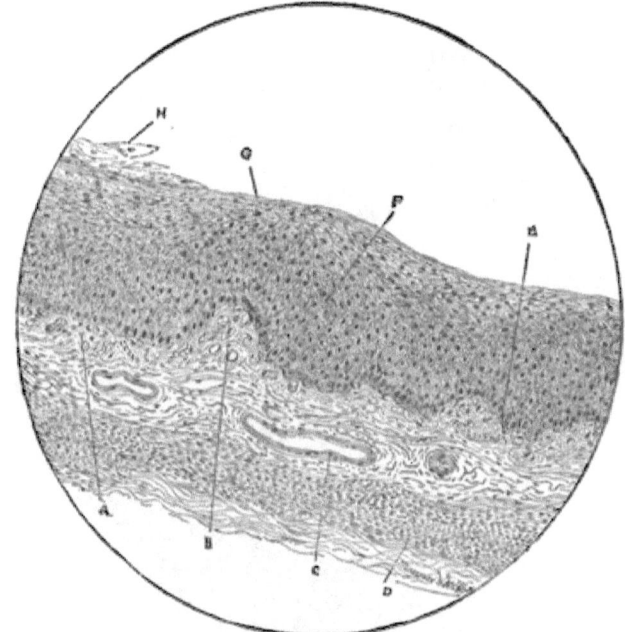

FIG. 94.—VERTICAL SECTION OF THE VAGINAL LINING AT PUBERTY. Stained with Hæma. and Eosin. × 400

    A. Sub-epithelial capillary plexus.
    B. Papillary arrangement of the mucosa.
    C. Large blood-vessels in the submucosa.
    D. Muscular wall of vagina.
    E. Deep cells of the lining epithelium.
    F. Middle strata of lining stellate cells.
    G. Surface cells in profile.
    H. Surface cells in plan—detached.

the same elements as they generally appear in a film of urine.) (h) The **nuclei**, evenly granular, usually larger than a red blood-corpuscle. (i) **Vacuolated cells**.

10. The **subepithelial vaginal structures**. (a) The large and abundant **capillaries** of the mucosa. (b) The **submucosa**,

not clearly separated from the superior coat, but easily recognized by the large vessels and the abundant connective tissue. (*c*) The **muscular vaginal wall**. (Here the muscular bundles are much better defined than in the uterine walls.)

PELVIS OF THE KIDNEY AND URETER.

TRANSVERSE SECTION OF THE URETER NEAR THE PELVIS OF THE KIDNEY, AND DETACHED CELLS FROM THE EPITHELIAL LINING OF PELVIS. (Figs. 95 and 97.)

(The arrangement and form of the cells lining the pelvis of the kidney and the ureters are precisely similar, and so far Fig. 95 will represent both.)

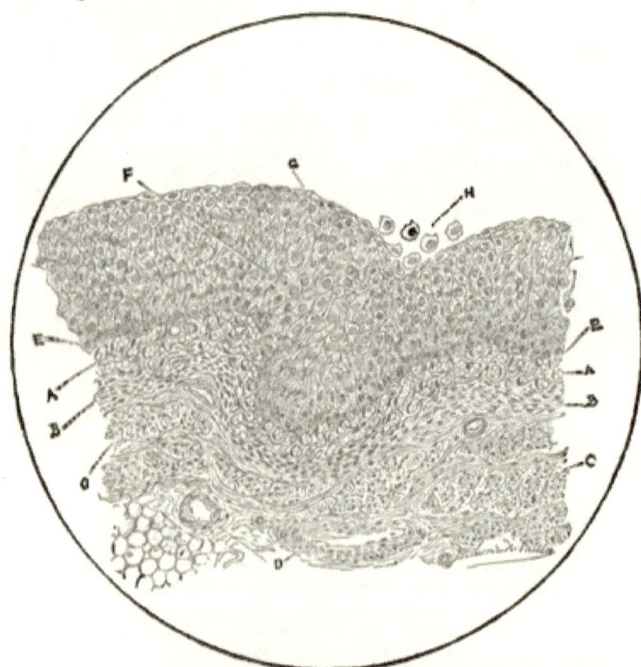

FIG. 95.—TRANSVERSE SECTION OF THE URETER, NEAR THE PELVIS OF THE KIDNEY. Stained with Hæma. and Eosin. × 400.

- A. Rich capillary plexus of the mucosa.
- B. Internal circular muscular coat.
- C. External longitudinal muscular bundles.
- D. Large vessels of the areolar adventitia.
- E. Deep layer of somewhat cubical cells.
- F. "Tailed cells" of the middle epithelial lining.
- G. Surface cells in profile.
- H. Surface cells in plan—detached.

OBSERVE:

(L.)

1. The **relative thickness of the epithelium.**
2. The **narrow mucosa.**
3. The **internal circular muscular belting.**
4. The **transversely divided** bundles of the **external** longitudinal **muscular layer.**
5. The **large arteries** between the muscular bundles.
6. **Adipose tissue,** more or less abundant in the loose cellular tissue surrounding these canals. (This element will afford a prominent feature of a section of the pelvis of the kidney, while the muscular tissue will be seen to a limited extent only.)

(H.)

7. **The epithelium.** (*a*) That it is of the **stratified** type, though poorly demonstrated. (*b*) The broad **basal attachment of the deep cells.** (*c*) The **elongate form of the cells** generally. (*d*) That the **borders are smooth** and **closely adherent,** unlike those of the vagina. (*e*) The more **flattened surface cells.** (*f*) The **outline of** the last, as seen in the detached specimens. (*g*) The very **large and finely granular nuclei.** (These cells contain peculiarly large nuclei, as compared with the size of the body. The deeper examples present tapering prolongations, generally at one end only, and are hence called "tailed cells." They will not be confounded with similarly shaped, though much larger cells from the bladder. The surface elements, while sometimes nearly circular, generally present one or two incurvations of the periphery, indicating their connection with the neighboring cells. These peculiarities are best exhibited in Fig. 97.)

8. (Review the objects previously examined with low power.)

THE URINARY BLADDER.

VERTICAL SECTION OF INNER PORTION OF WALL. (Fig. 96.)

OBSERVE:

(L.)

1. **The epithelial lining.** (*a*) That it is formed after the **stratified type.** (*b*) That, as compared with other previously studied portions of the genito-urinary tract, the **epithelium is thin.**

2. The thin mucosa and its small capillary supply.
3. The dense muscular portion, not arranged in bundles.

(H.)

4. **The epithelium.** (*a*) The **magnitude** of the cells. (*b*) The **broad-based cells without** any special **basement membrane.**

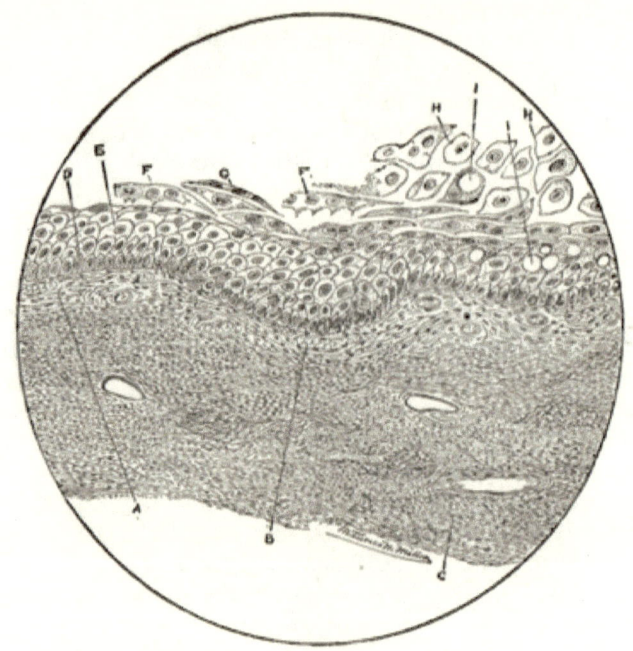

Fig. 96.—Vertical Section of the Lining Portion of the Bladder (Male) behind the Trigone.

Stained with Hæma. and Eosin. × 400.

- A. Connective tissue of sub-epithelial region, containing large amount of muscular fibre.
- B. Scant capillary supply of sub-epithelial region.
- C. Muscular wall of bladder.
- D. Large basement cells of the epithelial lining.
- E. Middle region of lining.
- F. Detached surface cells, showing processes beneath.
- G. Thin surface cells in profile.
- H. Squamous surface cells, seen detached, in plan.
- I. Vacuolated cells.

(*c*) The **three regions,** viz., **basal, middle,** and **superficial.** (*d*) The form of the middle cells; not unlike in outline, though larger, those of the corresponding region in the ureter. (This is best shown in Fig. 97.) (*e*) The large, scaly, and often **fused sur-**

face epithelia. (Note that while these, when seen in plan, all appear flat, it is only those of the extreme surface that are simple scales; the less superficial examples show, when viewed in profile,

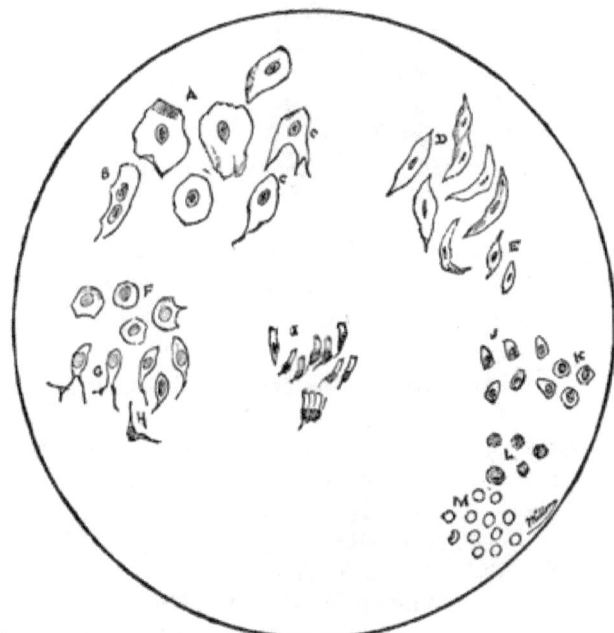

FIG. 97.—CELLS FROM GENITO-URINARY TRACT. ISOLATED BY TEASING, AS IN TEXT. × 400.
A. Surface bladder cells.
B. The same seen partly in profile.
C, C. Bladder cells from deeper layers.
D. Surface vaginal cells.
E. Vaginal cells from deeper layers.
F. Superficial cells from pelvis of the kidney or upper ureter.
G, H. From the same as F—deep layers. "Tailed cells." G. is the more usual form.
I. Ciliated vaginal cells.
J, K. Cells from collecting tubules of the kidney.
L. Pus-corpuscles. Not stained.
M. Red blood-corpuscles.
L and M are introduced as measurement standards of comparison.

prolongations from the under surface, by means of which union is effected with the deeper cells.) (*f*) **Vacuolated cells.** (These vacuolations do not occur in the basal layer.)

THE OVARY.

The ovary consists of a stroma or ground substance of connective and smooth muscular tissue, in which are scattered various sized spherical bodies or *Graafian follicles*.

The stroma is divided into three layers or regions, which are not very sharply differentiated.

The ovary is covered upon its free surface with a single layer of cells which in early life are cylindrical, becoming shortened with advancing age until after the *menopause*, when only flattened scales can be demonstrated.

Immediately beneath the epithelium a thin layer of fibrous tissue presents, with a free admixture of smooth muscle cells, and is termed the *tunica albuginea*.

The *cortex* proper, or second layer, is distinguished by the Graafian follicles, which will be described later.

The central portion of the organ, the *zona vasculosa*, is largely occupied by thick-walled blood-vessels, among which the extremely tortuous arteries are specially evident. Occasionally may be seen in this region somewhat ovoid nodules in varying degrees of retrograde change—the *copora lutea*. They present the phenomena resulting from the maturation of the follicle during menstruation. The accompanying illustration was drawn from a corpus luteum which had formed in the site of a Graafian follicle, the contents of which had escaped at some menstrual epoch, and been *followed by impregnation*.

PRACTICAL DEMONSTRATION.

The ovary of a young animal is to be preferred. If the organ cannot be obtained from the human subject, the female of almost any domestic animal will provide an excellent demonstration for the histological elements. Let the tissue be hardened with strong alcohol, and sections be cut vertically to the free surface and stained with hæma. and eosin. The sections should include at least one-half the depth of the organ, so as to exhibit all of the regions.

SECTION OF THE ADULT HUMAN OVARY. (Fig. 98.)

OBSERVE:

(L.)

1. The **tunica albuginea.** (Note that the layer is not of uniform thickness, and is composed largely of smooth muscular tissue,

THE OVARY. 145

as shown by the numerous elongate nuclei. Search particularly for and note the character of the epithelial covering.)

2. The **cortical layer**, containing numerous Graafian follicles, and possibly a corpus luteum. (Note the aggregation of the smaller follicles in the extreme outer portion of the region.)

3. The **zona vasculosa**. (Note the unusual thickness of the vascular walls and the irregular outline of section, on account of their tortuous course.)

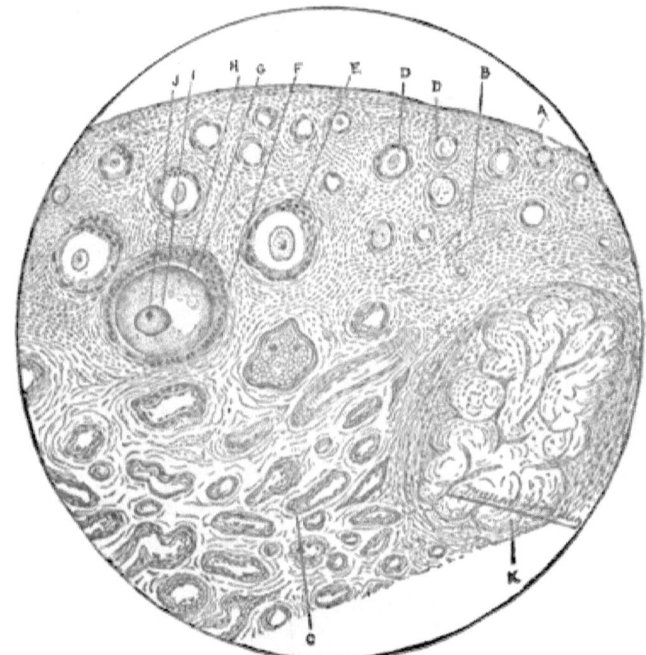

FIG. 98.—SECTION OF OVARY FROM A WOMAN 35 YEARS OLD. Stained with Hæma. and Eosin. × 250.

  A. Surface of the ovary.
  B. Muscular stroma.
  C. Large tortuous, thick-walled arteries of the central portion of the organ.
  D, D. Small Graafian follicles of the superficial zone
  E. Larger follicles of the deeper portion.
  F. Membrana propria of a Graafian follicle.
  G. Membrana granulosa of the follicle. The line leads to the discus proligerus.
  H. An ovum.
  I. Germinal vesicle.
  J. Germinal spot.
  K. An old corpus luteum.

(H.)

4. The **Graafian follicles**. (*a*) **Their diameter**, varying from one-one-hundredth to one-one-thousandth of an inch. (*b*)

The **membrana propria**. (This is difficult to separate from the stroma of the ovary itself, except in more mature follicles than shown in the section.) (c) The **membrana granulosa**. (This, in general, appears to be the outer layer of the follicle, on account of the difficulty of separating the membrana propria from the stroma proper of the ovary. Note that it is composed, in the smaller and less mature follicles, of pavement cells, and that the cells becomes thicker with maturation, until columnar cells in single layer result.) (d) **The ova**. (These are contained within the follicles, excepting they may have become detached during manipulation of the section, and occupy the greater area of the same.) (e) The **zona pellucida** (the thin wall of the ovum). (f) The **discus proligerus**. (This will be recognized as a mass of polyhedral cells, connecting the ova at one side with the columnar cells of the membrana granulosa. These cells will proliferate later in the development, and completely inclose the ovum.) (g) The **germinal vesicle**. (Contained within the ovum. The contents appear granular; it, as well as the ovum, is fibrillated; but this demonstration cannot be made excepting the animal be killed for the purpose, and the tissue elements fixed before changes, which quickly follow death, occur.) (h) The **germinal spot**. (Appearing as a small dot within the germinal vesicle. The ovum presents the characteristics of what it indeed is—a typical cell, with *wall, body, nucleus, and nucleolus.*)

5. The **corpus luteum**. (The example shown in the drawing, as I have already said, was developed after the contents of the Graafian follicle, which it represents, had suffered impregnation; and it has arrived at the later stage of the series of phenomena connected with its development—the stage of cicatrization. The cicatricial tissue, to which the letter K points, indicates the remains of the membrana granulosa. Outside is seen the thickened membrana propria, while among the contents will be found pigment-granules and fat-globules, imbedded in a structureless, gelatinous stroma. This material results from changes in the clot of blood effused after the escape of the ovum. I do not tabulate these elements, as it is extremely improbable that the student will find a corpus luteum in precisely the condition of the one represented until he has examined a large number of specimens.)

DEVELOPMENT OF THE OVUM.

As has been previously shown, the ovary is covered with epithelium; and singular as it may appear, the fifty thousand Graafian follicles, which it is estimated are developed during the life of the human female, have their origin in these cells.

During fœtal life, this surface epithelium undergoes a very rapid proliferation, and chains of cells are imbedded in the stroma of the ovary. A little later in the development, separate portions or links of these chains are cut off by the ingrowth of the stroma. The little groups of cells thus isolated become each a Graafian follicle.

Scattered among the columnar cells, larger, more nearly spherical cells are also found, the *primordial ova*. These are also imbedded in the substance, and one at least will always be found among the minute collections of cells which have been isolated.

In the process of development, each group of cells becomes a Graafian follicle with its contained ovum, the columnar cells forming the wall proper, and the primordial cell the ovum with its vesicle and germinal dot.

PRACTICAL DEMONSTRATION.

The ovary from a still-born babe is to be removed with the scissors, exercising the utmost care that the surface be not touched. Place immediately in strong alcohol, and in twenty-four hours it will be fit for cutting. Cut extremely thin sections at right angles to the free surface and including the same. Stain with hæma. and eosin. Mount in dammar.

OVARY OF HUMAN FŒTUS OF EIGHT MONTHS.
(Fig. 99.)

OBSERVE:

(L.)

1. **The free surface.** (Note the occasional depressions which mark the involution of epithelia.)

2. **The layers.** (Note the absence of demonstrable tunica albuginea and the great area occupied by the cortex. The vessels of the central portions are unlike the ovary of mature life; large, not numerous, and thin-walled.)

(H.)

3. The primordial ova of the surface epithelium.

4. The projecting lines or chains of epithelium undivided. (Here the cells seem rather elongate.)

5. Chains which are in process of subdivision.

6. Young Graafian follicles in columns at right angles to the surface of the ovary.

7. The discus proligerus, in many instances yet composed of flattened cells.

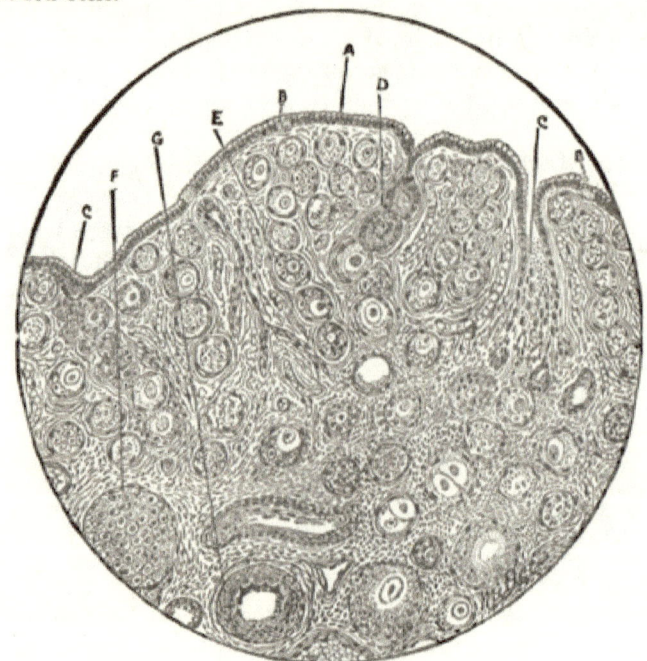

Fig. 99.—Section of Ovary of Child. Death Ten Days after Birth. × 350.

A. Germinal epithelium, covering surface of the ovary.
B. Primitive ova.
C, C. Projection of surface epithelium within the organ.
D. Constriction of the projected chain or cord of epithelium and isolation of portions to form Graafian follicles.
E. Chain of Graafian follicles. The stroma is seen filled with previously-formed follicles which have become now isolated.
F. A large Graafian follicle. It has been cut in half; the ovum has fallen out; and the membrana granulosa is seen lining the cup-shaped cavity.
G. Large arteries of the central portion of the ovary.

8. Follicles showing discus proligerus as columnar cells.

9. Follicles showing great proliferation of discus proligerus.

10. **Ova in early development** from primordial cells, with granular vesicle.

11. **Instances of development of two, possibly three, ova in a single follicle.**

12. **Large blood-capillary supply** of cortex, vessels generally parallel with the Graafian chains.

THE SUPRARENAL CAPSULE.

These bodies are attached by areolar tissue to the summit of the kidneys, and consist of several folia or leaflets. An examination of one of these leaves will give us an idea of the organ as a whole. The plan of structure seems to be as follows:

In the connective tissue which supports the folia are found arterial branches derived from the phrenic (and sometimes from the renal before it enters the kidney). These arteries penetrate the organ, break up immediately into capillaries, which finally converge toward the centre of the leaflet; the blood is here collected in thin-walled veins, by which it is drained into the suprarenal vein, thus leaving the capsule.

The capillary meshes vary in form and size, according to their position. Near the circumference of the leaves the meshes are small and ovoid, while, as the centre is approached, they become elongated. These spaces between the capillaries are filled with compressed, globular, nucleated cells, the smaller containing only perhaps six or eight, while the longer may be occupied by thirty or forty of these cell elements, which constitute the parenchyma of the organ. This variation in size of the cell compartments, contributing, as it does, to alter the appearance of the different zones of the tissue, has given rise to a division into *cortex* and *medulla*, with subdivisions even of these. There is no histological or physiological difference, as we believe, between the different parts of the folia of the suprarenal capsule, except as has been indicated. The structure is exceedingly simple, although its function is not settled beyond question.

PRACTICAL DEMONSTRATION.

The tissue is best hardened in strong alcohol, and should be cut as soon as the hardening is complete. It will be sufficiently firm to admit of the thinnest sections being made free-hand or with a simple microtome. The cuts, stained with hæma. and eosin, give excellent differentiation.

HUMAN SUPRARENAL CAPSULE. (Figs. 100 and 101.)

SECTION OF A SINGLE LEAFLET, CUT TRANSVERSELY TO THE CENTRAL VEINS. STAINED WITH HÆMA. AND EOSIN; MOUNTED IN DAMMAR.

OBSERVE:

(L.)

1. Section of **arterial twigs on the border of the leaflet**.
2. The **convergence of the parenchyma** toward the centre.

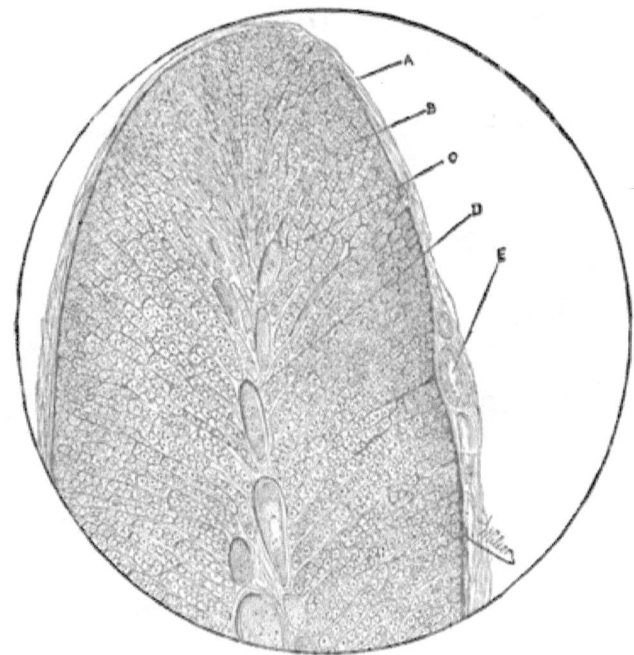

FIG. 100.—VERTICAL SECTION OF A SINGLE LEAFLET OF THE SUPRARENAL CAPSULE. Stained with Hæma. and Eosin. × 60.

A. Fibrous tissues surrounding and connecting the leaflets.
B. The outer portion, consisting of small compartments—the so-called cortex.
C. The central, elongated cell-compartments—medulla.
D. Large thin-walled central veins.
E. Arteries ramifying in the outer fibrous tissue which supply the parenchyma.

3. The **large and thin-walled central veins**.
4. The small **size of the parenchymatous areas on the outer borders and their elongation within**.

(H.) Fig. 101.

1. The **capillary plexus**, forming ovate or **elongate meshes.**
2. The compressed **globular cells of the parenchyma.** (Note that the cells are small in the small compartments, as though crowded. This is due, in a measure, to the contraction of the tissue from the rapid hardening.)

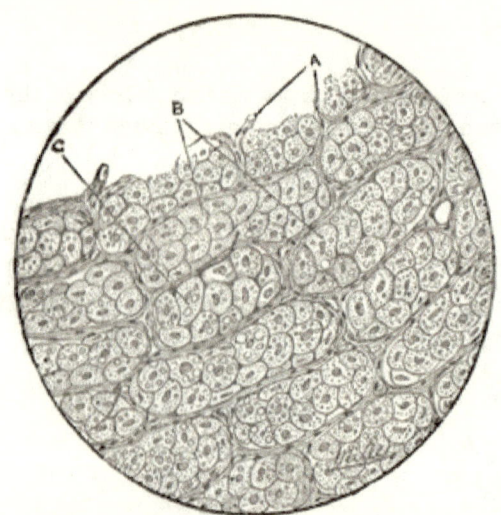

FIG. 101.—SAME SECTION AS FIG. 100, MORE HIGHLY AMPLIFIED. REGION MIDWAY BETWEEN THE FIBROUS INVESTMENT AND THE CENTRE OF THE LEAFLET. × 400.

A. Blood-capillaries, arising from the arteries seen in the preceding illustration; and ramifying in the connective-tissue framework.

B. Compartments—lobules—formed by delicate connective-tissue prolongations from the fibrous capsule.

C. Lobular parenchyma. These large somewhat rounded cells are generally mononucleated, contain fat-globules, and are frequently pigmented.

3. The minute **fat-globules** in the parenchyma. (This I believe to be physiological, and not unlike the fat-storing observed in the parenchyma of the liver.)
4. Yellow **pigment-granules** in the parenchyma.

THE SALIVARY GLANDS. PANCREAS. PLAN OF GLAND STRUCTURE.

GLANDS.

A gland is an organ—frequently subsidiary to and located within other organs—whose cells manufacture from the blood products to be utilized in the maintenance of physiological integrity.

Glands are tubes or cavities, with connective-tissue walls lined with cells of a columnar type. Around, and in close proximity to the lining, is spread a plexus of blood capillaries.

The essential parts of a gland are, therefore:

1. A *duct*, or efferent conduit for the secretion.
2. *Parenchyma*, cells engaged in secretion.
3. A *blood-vascular supply*.

TUBULAR GLANDS.

The simplest gland structure is presented in the form of a tube. Glands are, frequently, little more than tubular depressions in

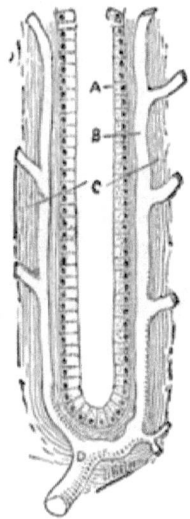

Fig. 102.—Diagram. Simple Tubular Gland.

A. Lining cells—parenchyma.
B. Capillary plexus, supplying the parenchyma.
C. Connective tissue supporting capillaries.
D. Arterial supply.

mucous surfaces. Examples are found in the uterus, stomach, small intestine, etc.

COILED TUBULAR GLANDS.

Tubular glands are often greatly elongated, with the blind extremity coiled. This variation presents the simplest differentiation between the part of the tube which is secretory, and the duct, or drainage part. With this change in function of the different extremities of the tube will occur a change of epithelium. The cells belonging to the duct-end will usually retain the columnar form; while the actively secreting elements will become enlarged,

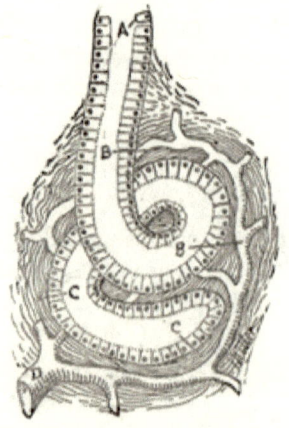

FIG. 103.—DIAGRAM. COILED TUBULAR GLAND.
Same references as Fig. 102.

more nearly filling the tube, and assume a polyhedral form from pressure.

Examples have already been seen in the sweat-tubes of the skin —sudoriferous glands—and the mucous glands of the submucosa of the larger bronchi.

BRANCHED TUBULAR GLANDS.

With the branching of the duct portion of gland tubules, there usually occurs a dilatation of the extremities into alveoli, although pure examples of branched tubular glands are afforded in the gastric and intestinal glands, those of the cervix uteri, etc.

The most nearly typical branching of gland-like tubules is afforded by the tubuli uriniferi of the kidney—although not contained in a true gland. The tubules here present other features peculiar to them, which will be referred to under the proper head.

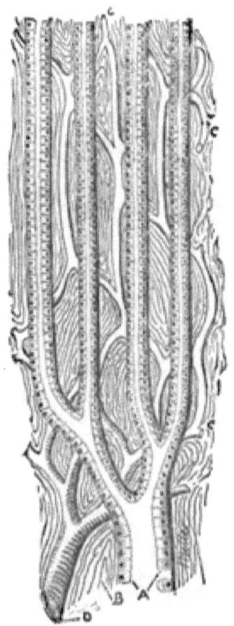

FIG. 104.—DIAGRAM. BRANCHED TUBULAR GLAND.
References same as Fig. 102.

ACINOUS GLANDS.

The dilatation of branching tubules, referred to under the previous heading, results in the formation of acinous glands. They are formed by the subdivision of a main tube or duct, with repeated branching of the secondary tubules. Collections of terminal branches often result in globular masses which are more or less perfectly isolated from one another by connective tissue. In this way *compound acini* are produced, such as the pancreas, the salivary, mammary, and buccal glands.

The acini may be developed into alveoli—as in the active mammary, and in the sebaceous glands. These are usually filled with polyhedral cells, or with the products of fatty degeneration of the same.

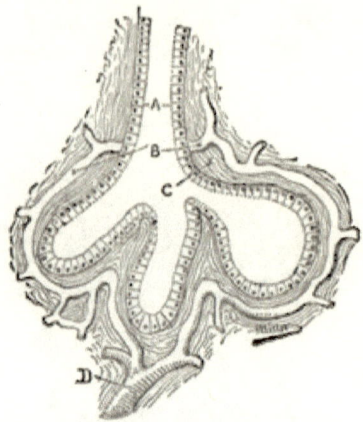

Fig. 105.—Diagram. Illustrating the Plan of Acinous Glands.
References same as Fig. 102.

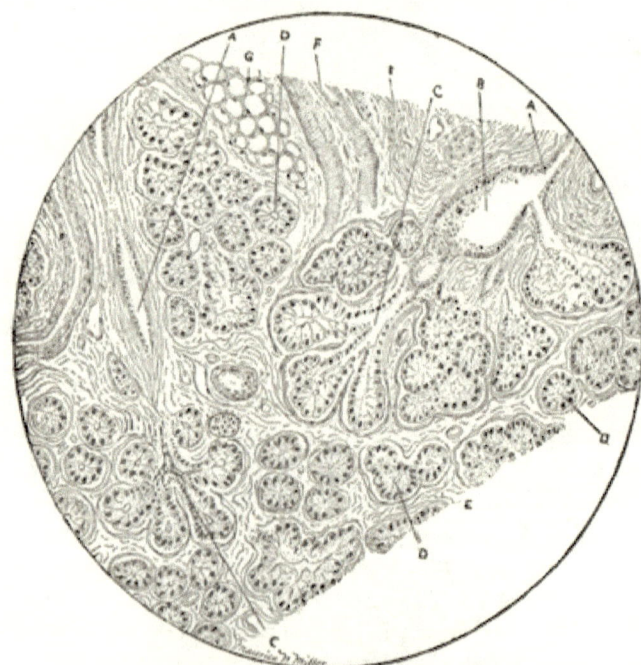

Fig. 106.—Section of a small portion of the Parotid Gland.
Stained with Hæma. and Eosin. × 250.

A. Narrowing of the duct from a small lobule, before entering a larger duct.
B. Dilatation of a duct after leaving a small lobule.
C. Primary lobules, in nearly L. S.
D. Acini in T. S., showing the minute lumen.
E. Connective tissue supporting the gland.
F. Striated muscular fibre adjacent to the gland.
G. Adipose tissue in the loose areolar tissue.

THE PAROTID GLAND.

The *Parotid, Submaxillary, Sublingual, and Buccal Salivary Glands* are typical glandular structures, with individual peculiarities only in respect to the cell elements; these vary according to the nature of the secretion formed in each.

The parotid is a *compound acinous gland*, leading from which is a principal duct—lined with tall columnar cells—which collects the fluid saliva from the different divisions of the organ.

As the duct penetrates the gland it branches freely, the lumina becoming smaller and the cells shorter as the deeper parts are approached.

Each terminal duct is in connection with several acini. The connective-tissue adventitia of the duct becomes the thin wall of

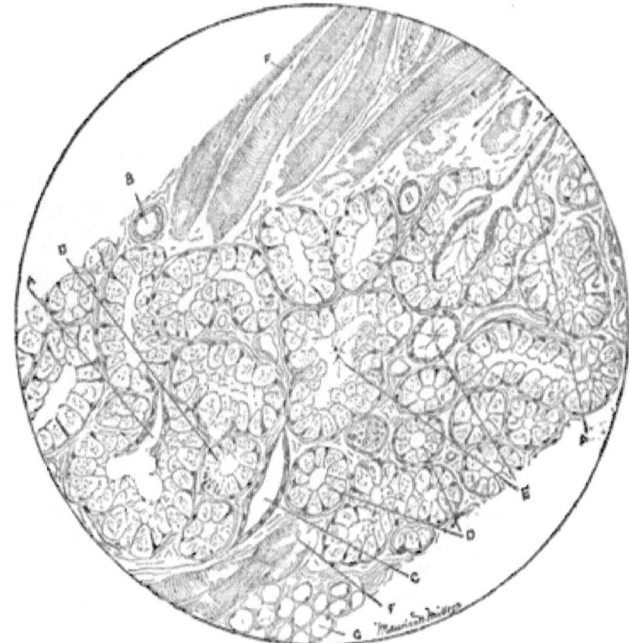

Fig. 107.—Section of part of the Submaxillary Gland. × 250.

    A. Narrow duct from terminal lobules.
    B. Small duct in T. S.
    C. Small duct in oblique section.
    D. Transversely-divided acini, showing large lumen.
    E. Mucous remaining in the lumina.
    F. Striated muscular fibres.
    G. Adipose tissue.

the acinus, and the lining cells broaden, frequently become polyhedral, and are bluntly pointed. The cells so nearly fill the acini as to leave a small and not easily recognized lumen.

The gland is richly supplied with blood-vessels.

THE SUBMAXILLARY GLAND.

The submaxillary is presented as an example of a typical *mucous gland*. As I have previously said, the general arrangement is not unlike that of the other salivary glands.

Similar structures are found in the submucosa of the mouth, tongue, fauces, trachea, and the larger bronchi.

Its peculiarity appears in the parenchyma, and will be noticed later.

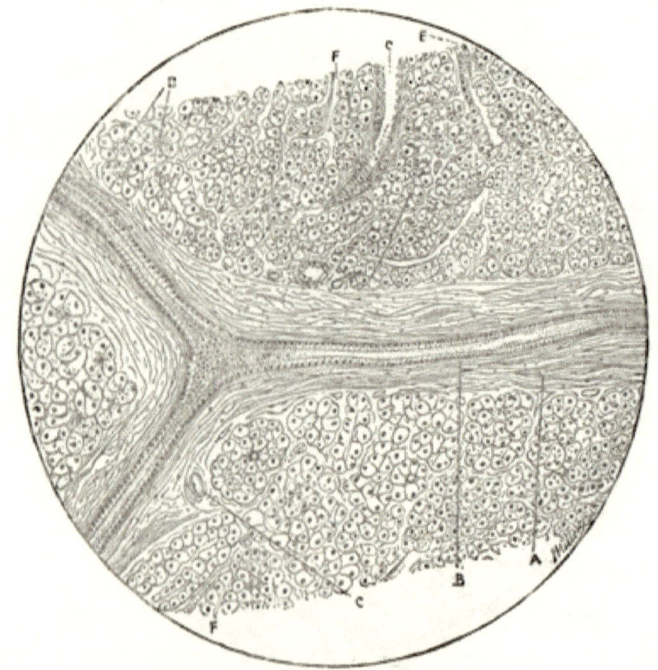

FIG. 108.—SECTION FROM THE PANCREAS.
A. Wall of a large duct.
B. The somewhat cubical lining cells.
C. Arteries.
D. Lumen of the acini, T S.
E. Terminal duct entering a lobule.
F. Acini in L. S.

THE PANCREAS.

The histology of the pancreas is, in general, that of a true *serous gland*—*e.g.*, the parotid. It has been called by physiologists the abdominal salivary gland.

The cells, constituting the parenchyma, are somewhat smaller; the lobules less regular in size and form; and the lumen of the acini much less easy of demonstration, in an ordinary hardened section, than the same in the parotid. The vascular supply is also more abundant.

The branches of the pancreatic duct are provided with a very thick adventitia, are lined with short columnar cells, and seldom present the dilatation, which generally occurs in a serous gland, on entering the lobule.

PRACTICAL DEMONSTRATION.

PAROTID AND SUBMAXILLARY GLANDS, AND THE PANCREAS.

The tissue must be fresh, divided in small pieces—not larger than a quarter of an inch cube—and hardened by placing in ninety-five per cent alcohol for twelve hours, after which fresh spirit should be substituted. If, after the lapse of another twelve hours, the tissue should not be sufficiently firm, it should be placed in a small quantity of absolute alcohol for three hours. Sections should be made immediately after hardening—as more prolonged action of the strong spirit will cause the tissue to contract.

Sections may be cut with or without a simple microtome—the desideratum being *thin rather than large cuts*.

Stain lightly with hæma. and deeply with eosin.

After sections of hardened tissue have been examined, the glandular parenchyma may be profitably studied in teasings from tissue which has been in Müller twenty-four hours. Wash the teasings on the slide with a liberal supply of water, removing the same from time to time with blotting-paper. Add a drop of hæma. solution; and, after washing this away, add a drop of glycerin, and cover. This method is very generally useful for teased or scraped fragments of glandular structures.

(Figs. 106, 107, and 108.)

OBSERVE:

(L.)

1. The **connective tissue**. (Most abundant in the parotid, and least so in pancreas.)

2. The **ducts**. (Note the **flattening of the lining columnar cells**, as the ducts approach the acini, until **mere scales** result.

Also the thick connective-tissue adventitia, especially demonstrable in the pancreas.)

3. The **lobules**. (These are formed by several acini, and are typical only in the parotid—at least they only here appear well formed. It must be remembered that only one plane is visible, and that there is little perspective.)

4. The **acini**. (Note the lumina—**large** in the submaxillary, **less so in the parotid**, and **least**, and often difficult to make out, **in the pancreas**.

5. The **blood-vessels, muscular and adipose tissue**. (The two latter are demonstrable only in the salivary glands, and do not belong properly to the gland itself. The capsule of the pancreas, in common with such structures in general, contains adipose. The abundant inter-acinous capillary plexuses of the pancreas require the high power for satisfactory demonstration.)

(H.)

6. The **parenchyma**. (*a*) The small but distinct **shortened columnar cells** of the acini **of the parotid**. (Observe that they are frequently so formed that the convexity of one cell fits into the concavity of its neighbor. Where seen in transverse section, the outline is a polygon. Note especially the **change in** the parenchymatous elements as the **terminal duct** merges into an acinus.)

(*b*) The **large, swollen cells of the mucous gland**—submaxillary. (Observe the comparatively **clearness** of the cells. They contain a very delicate **reticulum**, and their **nuclei** are often **obscured** and frequently seen to be **placed at the junction of the cells**.)

(*c*) The **rounded**, often polyhedral **cells of the pancreas**. (They resemble the parotid elements, although smaller and less granular.)

THE LYMPHATIC SYSTEM.

The Lymphatic System is a circulatory apparatus of exceedingly complicated arrangement. It comprises:

1. A system of *irregular clefts and cavities which are of almost universal distribution in the more solid tissues, in the framework and parenchyma of organs, around blood-vessels and viscera.*

2. *Nodules of sponge-like tissue,* improperly called lymphatic glands.

3. *Channels of communication,* consisting of *capillaries and ducts.*

4. A *central reservoir*—the *receptaculum chyli.*

5. *Large efferent ducts,* by means of which the contents of the system are, eventually, poured into the blood, in both sides of the neck at the junction of the internal jugular and subclavian vein.

6. A fluid, *lymph,* containing numerous nucleated bodies or *lymphoid cells,* and various substances in solution.

The whole provides a channel for the introduction of formed and nutrient elements into the blood; as well as affording drainage for the tissues, the products of which are also emptied into the blood-vascular system, to be afterward eliminated by special organs.

The circulating lymph always passes in a direction toward the venous system. This current is established in some of the lower animals by means of distinct, pulsating, hollow organs, or lymph hearts; but no corresponding structure exists in man, and the system becomes here subordinated to the blood-vascular apparatus.

In man, the maintenance of the lymph-flow is due largely to a negative pressure, consequent upon the connection between the termini of the lymph-vessels and the veins. Without doubt the pumping motion of the intestinal villi presents a factor in the establishment of a current in the lacteals toward the mesenteric vessels. The perivascular lymph receives an impetus with each cardiac systole. The muscular contractions of inspiration contribute motility to the contents of the diaphragmatic lymph-channels, in a direction against gravity. Indeed, the contractions of nearly every muscular fibre, whether skeletal or organic, lend their aid to lymph propulsion.

The direction of the lymph-current is determined by valves which resemble, somewhat, those of the veins.

Cavities lined with so-called serous membranes, may be considered as expanded lymph-channels.

LYMPH CHANNELS.

The larger and more regularly formed channels for lymph circulation, such as the mesenteric and thoracic ducts, do not differ, materially, in structure, from correspondingly sized veins. The irregular clefts in the interstices of fibrous tissues, serving as the primitive lymph-containing channels, have been already, and repeatedly, noticed. Fig. 109, although purely diagrammatic, will serve to show the relation of this system to the blood-vessels. A perivascular lymphatic channel is a sort of tubular investment of

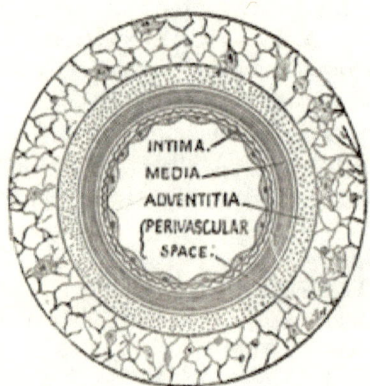

FIG. 109.—DIAGRAM. ARTERY IN TRANSVERSE SECTION, SHOWING THE PERIVASCULAR LYMPH-SPACE.

the blood-vessel, lined with flattened endothelia sending prolongations inward; these prolongations branch, and are finally in communication with a layer of cells covering the adventitia. In this manner, in close apposition to parts of the vascular system, a system of channels is provided, *within which the lymph may slowly percolate.*

The largest lymphatic channels in the human body are the cavities of the peritoneum and pleuræ. They are in connection one with the other, and with the lymphatic system generally; and these channels of communication between the great abdominal and thoracic lymphatic cavities present, perhaps, as the most convenient and typical for demonstration.

PRACTICAL DEMONSTRATION.

LYMPHATIC VESSELS OF THE CENTRAL TENDON OF THE DIAPHRAGM.

(Figs. 110 and 111.)

This demonstration had best be made with tissue from the rabbit, inasmuch as the slightest decomposition of the epithelium would be fatal to success.

A small (preferably white) rabbit should be quickly killed by decapitation, and immediately suspended by the hind legs, so as to thoroughly drain the body of blood. As soon as the blood has ceased dripping, open the thoracic cavity by slitting up the skin along the median line, pushing it to the sides and removing the sternum. In this operation, work rapidly and avoid soiling the internal parts. Then with the fingers of one hand raise the lungs and heart from the diaphragm, and with a large camel's-hair brush proceed to quickly, and quite forcibly, pencil the white glistening surface of the central diaphragmatic tendon, moistening the brush from time to time in the lymph of the pleural cavity. Should the quantity of fluid be small, add a little distilled or previously-boiled and filtered water. The object of the brushing is to remove the epithelial cells which cover the surface, and which would otherwise hide the lymph-spaces. After the pencilling, drain away the fluid, and pour over the brushed surface a solution of one grain of nitrate of silver to an ounce of distilled water.* Allow the silver solution to remain for twenty minutes in contact with the tissue, the body meanwhile being kept away from the bright sunlight; then pour off the solution, wash the surface twice with distilled water, and afterward allow water from the tap to flow over the parts for at least five minutes.

If you observe the directions carefully, the surface of the tendon will lose its original glistening appearance and become whitish and opaque.

The tendon, or such portion of it as you wish to preserve, may be cut out with the scissors after the washing, thrown into glycerin, and placed in the sunlight until the surface becomes brown. With the forceps tear off small pieces of the stained side, say one-half inch square, and examine in glycerin, or mount them permanently in the same medium.

The demonstration of the channels of the lymphatic system is based upon the following:

1. *Lymph-channels are always*, however small or irregular, *lined with flattened cells in a single layer*—i.e., pavement endothelium.

2. *The lining cells are cemented together with an albuminous substance.*

---

* Water which has been well boiled in a clean vessel, and afterward carefully filtered, may be generally employed in histological work when distilled water is not available.

3. *Nitrate of silver combines with the cement, forming albuminate of silver, which becomes dark brown when exposed to light.*

If you have been successful, the silver will have penetrated the tendon, and mapped out the lymph-channels, *indicating an outline of every lining cell by means of a dark border.* Failure will result only from non-attention to cleanliness in the handling of the tissue; the silver in which case becomes deposited generally over the surface. The margins or outlines of the cells, it must be remembered, are stained with the silver. The nuclei may be demonstrated by after-staining with dilute hæma., or better, borax-carmine. The mounting may be done in dammar, although the elastic fibres, of which the matrix of the tendon is composed, will become stiff during immersion, and show a tendency to curl and contract. If glycerin be used after carmine-staining, tissues should be washed thoroughly in water, subsequently to the oxalic-acid bath, transferred to equal parts of glycerin and water, and allowed to remain for an hour, at least, before mounting.

CENTRAL TENDON OF THE DIAPHRAGM. SILVER-STAINING. (*Vide* Figs. 110, 111.)

OBSERVE:

(L.)

1. The **division of the specimen into dark and light areas.** (The dark areas represent the more solid portions of the tissue or the partitions between the channels, and the light spaces are the lymph paths.)

2. The **lymph paths**—the light spaces. (These show, with this amplification, as irregular, winding, and anastomosing courses, marked with very delicate lace-like tracery—the silver lines.)

3. **Valves of the lymph paths.** (At points, the paths will be crossed by dark curved lines. These are imperfect valves, not unlike a single cusp of an aortic valve.)

(H.)

4. **Outlines of the cells lining the larger excavations** (lymph paths) in the tissue. (Note that the cells are generally elongate in the direction of the lymph path. The edges are frequently serrated.)

5. **Stomata,** minute openings at the junction of several cells.

6. The **construction of the valves.** (These are curved against the lymph flow, and covered with cells like other parts of the

channel. Note the change in form of the cells approaching and covering the valves.)

7. **Elastic fibres of the more solid parts of the tendon.**

8. **Lymph capillaries.** These will be seen in the partitions between the larger paths. In places they may be observed emptying into the paths, and again will appear as simple cavities, according to the manner sectioned.)

9. The **deeper capillaries.** (Careful focussing the portions

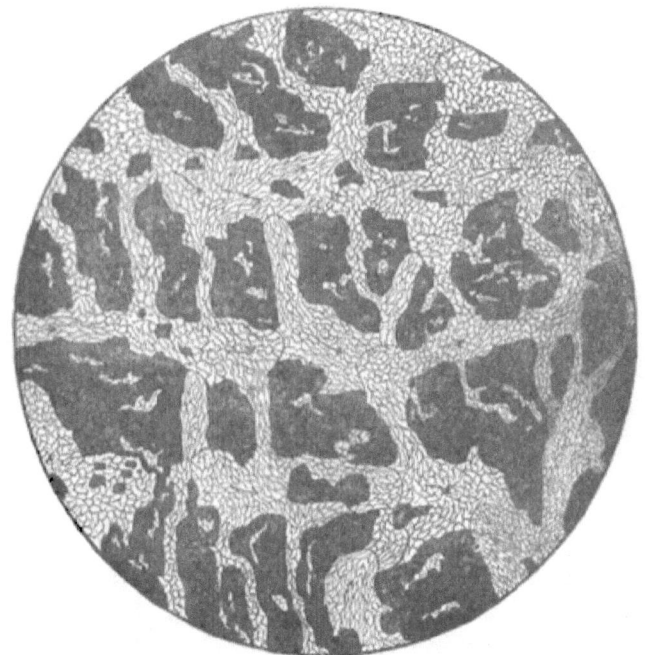

FIG. 110.—LYMPH-CHANNELS. CENTRAL TENDON OF DIAPHRAGM OF RABBIT. SILVER STAINING. × 60.

The dark portions represent the more solid portions of the tissue.
The light areas are the lymph-channels; and the direction of the flow is shown by the arrows.
The minute lines in the lymph-spaces are the silver-stained cement boundaries of the pavement cells lining the channels.
The valves appear as curved lines in the lymph-spaces.

of the tendon which appear most solid will reveal minute cell-lined channels or capillaries. The student must remember that we cannot penetrate tissues with the microscope to any considerable depth, but are restricted to nearly a single plane. If it were possible to penetrate with the eye the entire thickness of the tendon, we might trace the lymph paths or channels from the abdominal to the thoracic surface.)

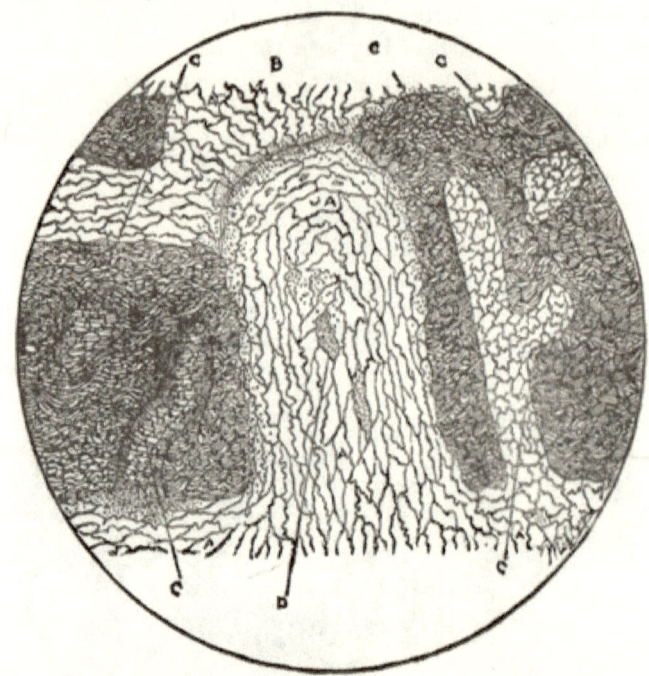

Fig. 111.—A Small Portion of Specimen shown in Fig. 110, more Highly Magnified.
× 350.

A, A, A. Large lymph-channel.
B. Valve in the course of last.
C, C, C. Lymph-capillaries in the more solid parts of the tendon.
D. Pavement cells upon which a large amount of silver has deposited. Failure to follow the instructions for the staining frequently results in a like deposition of silver over the whole surface.

LYMPHATIC NODES OR GLANDS.

At numerous points along the course of lymphatic vessels they penetrate small nodules of so-called *adenoid tissue*, which have been termed lymphatic *glands*. They are frequently microscopic; others, again, not unusually attain the size of a large pea. *They secrete nothing, hence are not glands.* They are somewhat sponge-like in structure, and the lymph filters slowly through them.

Most frequently several ducts enter one of these larger nodes, while perhaps only a single efferent will be found.

The histology of a lymph node is not always easily comprehended by the student, and I have endeavored to make a diagram (Fig. 112) that would simplify the matter somewhat. They are enveloped by a *capsule* of connective and involuntary muscular tissue, which sends *trabeculæ* into the body of the organ, and these branching posts support the structure as a framework. The interstices are quite small in the more central portion and larger toward the periphery; this has resulted in the application of the terms *medullary* and *cortical* to the respective parts. The nutrient blood-vessels are contained in the framework. *The compartments contain the structure peculiar to the lymphatic system*—viz., *adenoid tissue.*

Adenoid tissue consists of a mass of flattened cells, with numerous delicate fibrillar prolongations, which branch and anastomose so as to form an interwoven structure—*the adenoid reticulum.* Klein regards the cells as forming no essential part of the structure, but considers them as flattened plates attached to the fibrils. The meshes of the adenoid reticulum are in connection with the fibres of the trabeculæ and, with exception of the portion next the latter, are filled—crowded, in fact—with countless small spherical lymphoid cells. Those portions of the tissue which contain the cells are termed *follicular cords.*

The *lymph path* is the portion between the fibrous trabeculæ and the follicular cords.

When we learn that the trabeculæ, follicular cords, and lymph paths each pursue very tortuous and branching routes, we can appreciate the complexity of the organ as a whole.

The blood-vessel arrangement presents no anomalies. The small arterial trunks enter within the trabeculæ, finally break into capillaries which supply the follicular cords, etc., and the blood is then collected by the venules for the efferent veins.

Small diffuse collections of adenoid tissue have already been seen in many organs. These do not differ essentially from the tissue just described, excepting that there is no definite arrangement of trabeculæ and lymph paths, as in the compound lymph node; the lymph simply *filters through the reticulum,* the same being a part of the lymph-channel system of the tissue in which the adenoid structure may occur.

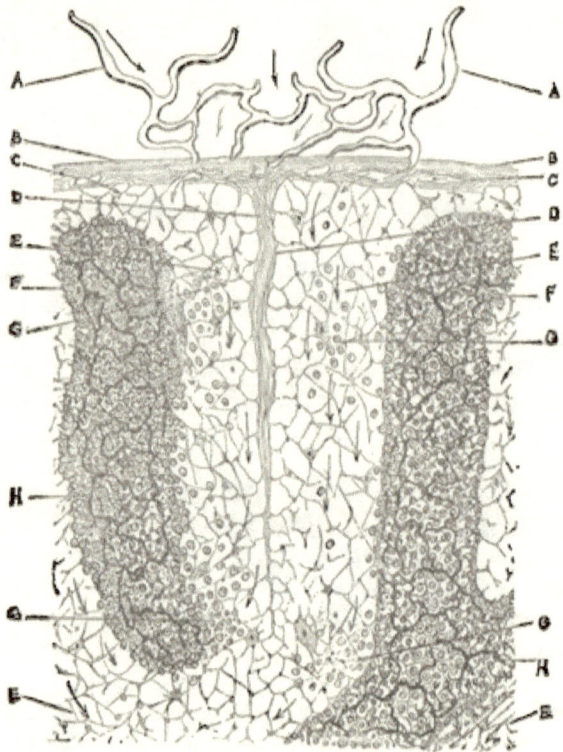

FIG. 112.—DIAGRAM. PERIPHERAL PORTION OF A LYMPH-NODE.
    A, A. Afferent lymph-vessels.
    B. Capsule of the node, with lymph-spaces C. C.
    D. Trabecula of connective tissue.
    E, E, E. Lymph path in the node.
    F, F. Follicular cords.
    G, G, G. Lymphoid cells in the cell network of the paths.
    H, H. Blood-capillaries of the cords.
        The arrows show course of lymph.

PRACTICAL DEMONSTRATION.

The *mesenteric lymphatic nodes* present the most typical structure, and may be obtained from the human subject, if fresh, although those from the dog are preferable, on account of the better condition of the tissue as usually secured.

The nodes should be sliced in half, placed in Müller for a week, and then hardened by two days' immersion in strong alcohol.

Sections should be mounted, of two kinds, viz., those including the whole area of the node—which need not be very thin—for demonstration of the scheme or plan of structure, and exceedingly thin ones, even though they may include only a small part of the organ, for study of the details of the adenoid reticulum. The latter purpose will be subserved by shaking a number of thin cuts in a test-tube with alcohol for a few minutes, and with considerable violence, even to the sacrificing of most of the sections. The agitation will dislodge the lymph cells, which otherwise would obscure the histology of the follicular cords.

Stain deeply, with hæma. and eosin, and mount the thicker sections in dammar, and those especially thin in glycerin.

SECTION OF MESENTERIC LYMPHATIC NODE.

(Figs. 113 and 114.)

OBSERVE:

(L.)

1. The **fibrous capsule.** (Note the elongate dots in the deeper parts of the capsule—the **nuclei of the smooth muscular tissue,** the **thick-walled arteries,** the **lymph spaces.**)

2. The **trabeculæ.** (Trace these as they penetrate the organ and observe that they frequently end abruptly, on account of having curved, so as to leave the plane occupied by the section. The **trabeculæ are not partitions,** like the interlobular pulmonary septa or the prolongations from the capsule of Glisson in the liver; they are not unlike rods or posts, making a framework and not producing alveoli. Find one divided transversely.)

3. The **follicular cords.** (They are recognized as granular masses between the trabeculæ. Observe the **varying forms,** largest and more spherical or ellipsodial, **near the periphery**—cortex. The smaller ones of the **central region** (medulla) must not be overlooked, as the differentiation is sufficiently marked between them and the variously sectioned trabeculæ.)

4. The **lymph paths.** (These can be appreciated by remem-

bering that the follicular cords do not entirely fill the spaces between the trabeculæ, and that the area between the two—*i.e.*, outside the cords—is the more open in texture, and contains the filtering lymph. They are more distinct in the cortex.)

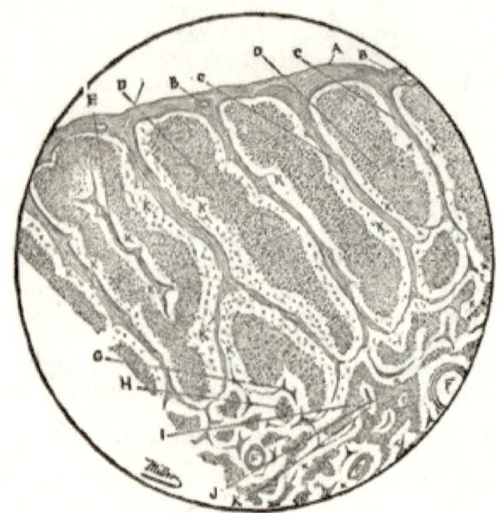

Fig. 113.—Vertical Section of a Lymph-node from the Mesentery. × 66.

    A. Capsule of node.
    B. Lymph-spaces in the last.
    C, C. Trabeculæ, L. S.
    D, D. Follicular cords, L. S.
    E. Obliquely sectioned trabecula.
    F, F. Large blood-vessels of the central portion of the node.
    G. Trabecula in T. S.
    H. Follicular cord in T. S.
    I. Small and irregular cords of the centre of the node.
    J. Obliquely sectioned trabecula of the centre of the node.
    K, K, K. Lymph-paths.

(**H.**)

5. The **histology of the capsule.** (*a*) The closely-united connective tissue with the scattering **elastic** fibres of the **external layer.** (*b*) The **smooth muscle of the deeper portions.** (*c*) Sections of **arteries.** (These may present of considerable size.) (*d*) The **lymph spaces.** (The differentiation is by the **flattened endothelia** of spaces which otherwise would be supposed mere rifts in the tissue, inasmuch as no definite or special wall can be detected.)

6. The **structural elements of the trabeculæ.** (They are similar to those of the capsule, excepting the elastic element, which

cannot here be demonstrated. Note the variously sectioned small arteries.)

7. The **follicular cords**. (In the thicker section, the field will be completely crowded with lymphoid cells. Select a thin field and observe: (*a*) The **lymphoid cells**. (These will be found varying in size from a very small red blood-disc to that of a large white corpuscle; some are filled with granules only, and others with one, two, and even three nuclei.) (*b*) The **branching endothelioid cells**. (*c*) The delicate fibrillæ of the **adenoid reticu-**

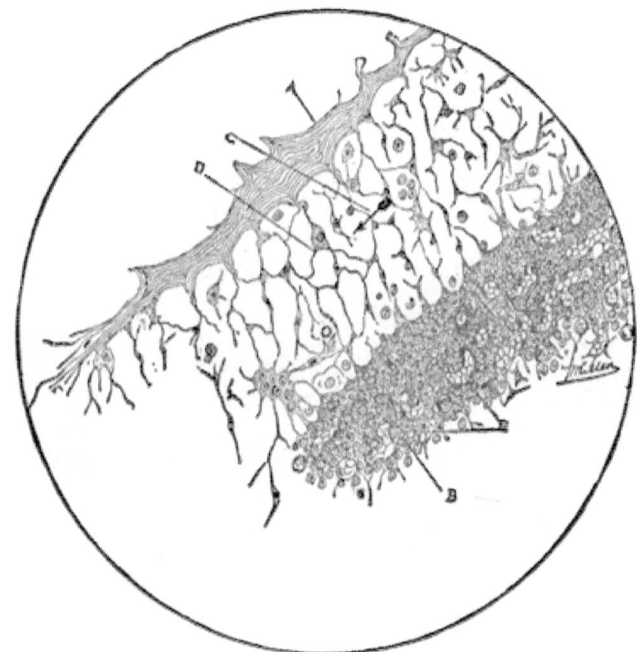

FIG. 114.—FRAGMENT OF SECTION SHOWN IN FIG. 113. MORE HIGHLY MAGNIFIED. × 350.
  A. Trabecula.
  B. Follicular cord.
  C. Lymph-path.
  D. Large branching cells of the path network.
  E. Capillaries of the cord.

lum. (You may endeavor to determine whether this reticulum exists as an offshoot of the endothelioid cells, or whether the latter are simply adherent to the broadened plates of the former.)

8. The **reticulum of the lymph paths**. (Observe that this is precisely like the reticulum of the follicular cords, as demonstrable

after shaking out most of the lymph corpuscles of the last.) (*a*) The **connection between the fibrillæ of the paths and those of the trabeculæ**.

9. **Capillaries of the paths and cords.** (These will be recognizable only by the regular succession of the contained red blood-corpuscles.)

THE SPLEEN.

The spleen presents no regular subdivision of parts which may be studied separately and combined afterward, as we are able to do with organs like the lung, liver, etc. The spleen is a *ductless* organ or so-called gland, and the plan or scheme may, perhaps, be best comprehended by following the blood distribution.

The splenic artery enters the organ, supported by a considerable amount of connective tissue, and rapidly breaks into smaller branches, from which the arterioles leave at right angles. The arterioles quickly merge into capillaries, which form plexuses through-

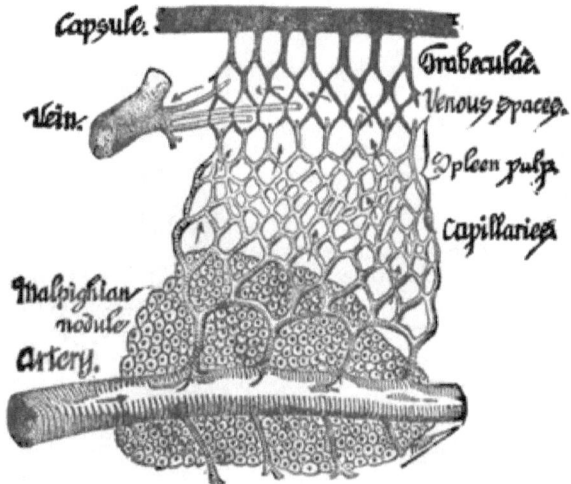

FIG. 115.—DIAGRAM. SHOWING THE COURSE OF BLOOD IN THE SPLEEN.

out the different portions of the organ. Here we meet with an anomalous structure.

The capillaries, instead of uniting to form venules as in the usual vascular plan, *empty their contents into small chambers or sponge-like cavities—the venous spaces*. The blood, after filtering through these venous interstices, is collected in larger, irregular, vein-like channels, which finally conduct the blood into the veins proper and out of the spleen.

The tissue, containing the vascular arrangement described in the last paragraph, is called *spleen pulp*.

The fibrous capsule which envelops the spleen sends trabeculæ within, which form a framework; and from this fibrils are sent off which branch, broaden, and inosculate *to form the venous chambers of the pulp.*

The arteries are frequently surrounded by nodules of adenoid tissue, sometimes globular, more frequently considerably elongated, and following the vessel for a considerable distance. These nodules are called *Malpighian bodies.* They bear no resemblance to similarly named structures in the kidney, excepting, perhaps, when seen in transverse section by the naked eye.

The spleen will thus be seen to consist of a fibrous trabeculated *framework,* the *pulp, blood-vessels,* and more or less isolated nodules of *adenoid tissue.*

PRACTICAL DEMONSTRATION.

The organ must be absolutely free from decomposition. If human tissue cannot be obtained in good condition, recourse may be had to the ox, which will provide an excellent substitute. The small supernumerary spleens, not infrequently found during *post-mortem* work, are most desirable, as sections can be easily made through the entire organ.

Pieces of tissue half an inch cube, including a portion of the capsule, should be hardened as directed for lymph nodes. Sections are easily made without the microtome, as the mass is very firm; they should be thin and stained with borax-carmine, and mounted in dammar or in glycerin.

SECTION OF HUMAN SPLEEN, CUT AT RIGHT ANGLES TO AND INCLUDING THE CAPSULE. (Fig. 116.)

OBSERVE:

(L.)

1. The **fibrous capsule**. (*a*) Its division into two **very distinct portions** or layers. (*b*) The clear **translucent appearance** of the tissue (**elastic**) of the **outer layer.** (*c*) The **darker deep layer** with elongate nuclei. (The elastic element of the capsule not infrequently becomes, in the human subject, considerably increased; and this development occurs irregularly, sometimes in the form of minute nodules. I do not know that they present any pathological significance.)

2. The **trabeculæ**. (The depth to which they may be traced will depend largely upon the direction of the section.) (*a*) That **these are not bands**, but bundles, more or less circular, in trans-

verse section. (*b*) Their **irregular course,** quickly **after leaving the surface.** (*c*) That occasionally a **small artery** may be found within them, though they are usually destitute of large vessels. (*d*) The elongate nuclei of the **muscular fibre** largely forming the trabeculæ.

3. The **large blood-vessels.** (*a*) The **arteries** more frequent

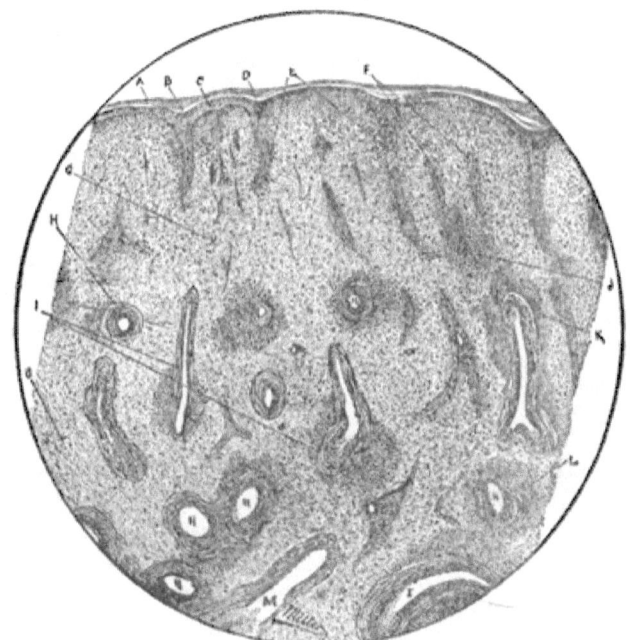

FIG. 116.—SECTION OF THE SPLEEN. × 60.
A. Elastic portion of the capsule.
B. Lymph-spaces of last.
C. Involuntary muscular portion of capsule.
D. Deeply pigmented portions of capsule.
E, E. Trabeculæ from C.
F. Trabeculæ in oblique section.
G, G. The spleen pulp.
H, H. Large arteries in T. S.
I. Arteries in L. S.
J. Adenoid nodule, not connected with an artery.
K. Adenoid nodule.—Malpighian body—along course of artery.
L. Adenoid nodule in T. S.
M. Vein.

than **veins.** (*b*) Their very prominent **adventitia.** (*c*) Their **tortuous** course.

4. The **adenoid tissue.** (This you will be enabled to recognize by the great number of lymphoid cells of the adenoid struc-

ture, the nuclei of which become stained very deeply blue with hæma., giving a very distinct differentiation. At this point, examine every part of the specimen, and endeavor to detect even the most minute collection of this tissue.) (*a*) Around arteries, constituting the so-called **Malpighian bodies**. (*b*) **Transverse sections of Malpighian bodies**, noting that the **vessel is seldom in the centre of the nodule**. (*c*) Nearly **longitudinal sections of Malpighian nodules**, observing that the adenoid tissue usually follows or surrounds the artery for a short distance only. (*d*) That the distribution is not confined to the arteries, but is quite common **around trabeculæ** and **beneath the capsule**.

5. The **Spleen-pulp**. (This will be found in those portions of the section not occupied by structures previously demonstrated; and will be determined by its light color. Review the whole area, and endeavor to differentiate every portion of the adenoid and pulp tissue. The staining will have been your principal guide thus far, the pulp elements appearing in strong contrast by their pink eosin color.)

(H.)

6. The **structural elements of the capsule**. (*a*) The **numerous minute lymph-spaces** and the **imperfect vascular supply**. (*b*) The nuclei of the **peritoneal cell covering**. (This presupposes that the section has been selected so as to include the peritoneal investment.) (*c*) The abundant and closely-packed **elastic fibrillæ**. (*d*) The **muscle nuclei** of the deeper parts. (*e*) Cells containing granular yellow **pigment**. (The quantity varies largely with different specimens.)

7. The **Malpighian nodules**. (*a*) The **arterioles**—very small and apt to escape attention unless filled with blood-corpuscles. (*b*) The **adenoid reticulum**. (This will be difficult of satisfactory demonstration, excepting the section be thin.)

8. The **elements of the pulp**. (*a*) **Large flattened cells**, the branches forming the meshwork of venous channels. (These are only susceptible of very satisfactory demonstration in the spleen of *leucocythæmia*.) (*b*) **Red blood-corpuscles**. Very numerous and often broken and distorted. (*c*) **Blood pigment**. (*d*) **Lymphoid or white blood-corpuscles**.

THYMUS BODY.

The thymus body (frequently and improperly called a gland) is an adjunct to the lymphatic system of—in man—fœtal and infantile life; disappearing, by an atrophic process, at or before the age of puberty.

It is enveloped by a fibrous capsule, partitions from which subdivide the organ into lobes and lobules. The lobules are generally subdivided into follicles, which are irregularly sized and shaped, while tending to an ovoid form.

It is in connection with the general lymphatic system by peripheral, afferent lymph-channels; and by efferent vessels which emerge from the hili of the lobes—the lymph having meanwhile traversed the mesh-like structure of adenoid tissue composing the follicles.

The blood-vascular system is in the form of a nutritive supply; the larger vessels occupying the fibrous framework, and sending branches into the follicles. The capillary plexuses are more abundant in the peripheral portion of the follicles. The blood is collected in the venous channels of the central or medullary area, and emerges from the organ by the veins which accompany the efferent lymphatics.

PRACTICAL DEMONSTRATION.

The organ should be obtained from a still-born infant, divided in small pieces, and hardened rapidly in strong alcohol. Sections may include an entire lobe, and be stained with hæma. and eosin.

SECTION OF THE THYMUS BODY FROM AN INFANT AFTER DEATH ON THE SIXTEENTH DAY.

(Fig. 117.)

OBSERVE:

(**L.**)

1. The fibrous capsule.

2. **Division by prolongations** of 1 into somewhat spherical lobes.

3. **Subdivision** of 2 into lobules.

4. **Subdivision** of 3 **into follicles.** (Note that these are not uniformly outlined by the connective tissue.)

5. The **subdivision of the follicles** into an outer, deeply-stained **cortex**, which completely surrounds a light centre, the **medulla**.

6. The **larger lymph-spaces** and **arteries** of the capsular and trabecular tissue.

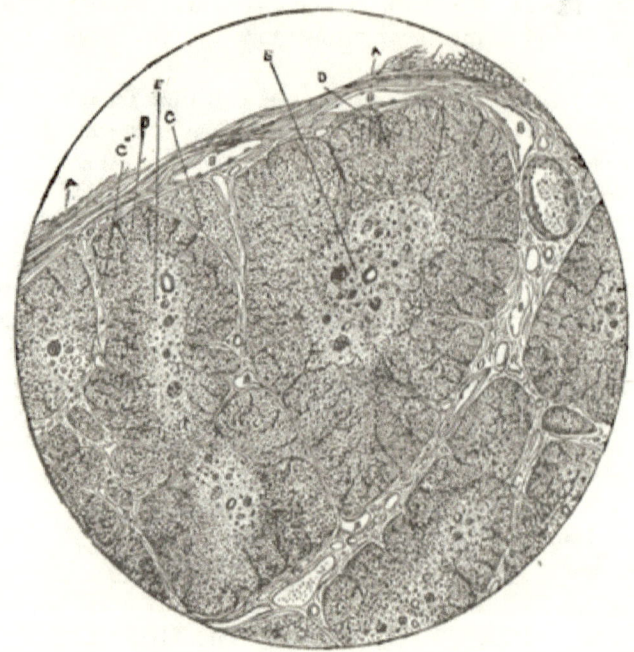

Fig. 117.—Section of a Portion of the Thymus Body, from a Child, Sixteen Days after Birth. × 60.

A, A. Capsule which divides the organ into lobes. Portions of six lobes are visible in the section.
B, B. Lymph-spaces.
C, C. Trabeculæ dividing the lobes into imperfect lobules.
D, D. Subdivisions of the last into follicles.
E, E. Central light portion of the lobules.

(H.)

7. The **cortex of the follicles**. (*a*) The numerous deeply-stained **lymph-corpuscles**. (*b*) The network of the **adenoid tissue**. (This will be greatly obscured by the lymphoid cells.) (*c*) The **blood capillaries**. Only recognized by the contained corpuscles. (*d*) Minute **trabeculæ** of connective tissue projected from the capsule.

8. The **medulla of the follicles**. (*a*) The **sparsity of lymph-corpuscles** as compared with the cortical portions. (*b*) **Large**

mononucleated cells. (*c*) Still **larger multinucleated cells.** (*d*) Larger—though varying in size—spherical bodies, **Hassall's corpuscles.** (These are composed of epithelioid cells, arranged concentrically, and are unlike any other structure found in the normal tissues of the body. They resemble very closely the smaller "brood nests" of epithelial cancer.) (*e*) Small thin-walled **venules.**

THE NERVOUS SYSTEM.

STRUCTURAL ELEMENTS.

The elements of the nervous system are:
1. Nerve Fibres.
2. Nerve Cells.
3. Connective Tissue.
4. Peripheral Termini.

NERVE FIBRES.

A typical nerve fibre consists of three portions, viz.: a central conducting portion, the *axis cylinder*; the medullary sheath, or *white substance of Schwann*; and the enveloping connective-tissue substance, the *neurilemma*. This constitutes a medullated nerve fibre, and is found largely in the trunks of the cerebro-spinal system. The trunks of the sympathetic system are composed princi-

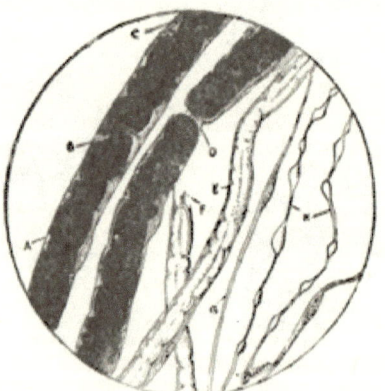

FIG. 118.—SEPARATED NERVE FIBRES. × 400.

A. Neurilemma of a fibre.
B. White substance of Schwann, stained with osmic acid, which hides the axis cylinder.
C. Nucleus of the neurilemma.
D. One of Ranvier's nodes in an osmic-acid stained fibre showing the axis cylinder between the separated portions of Schwann's sheath.
E. A medullated fibre, teased in normal salt solution. The medullary substance has become coagulated on exposure and removal. The axis cylinder is faintly seen.
F. Axis cylinder at torn extremity.
G. Non-medullated fibre.
H. Fibres without neurilemma. Small clusters of medullated substance are seen covering the axis at irregular intervals.

pally of fibres destitute of the white substance of Schwann—non-medullated nerves; while fibres minus the neurilemma exist in the trunks belonging to some organs of special sense.

After treatment with reagents, the axis cylinder (one-two-thousand-five-hundredth to one-fifteen-thousandth of an inch) may be split up longitudinally, and is found to be composed of fine (one-twenty-five-thousandth of an inch) primitive or ultimate fibrillæ, which present minute varicosities or swellings at irregular intervals.

The white substance of Schwann presents under the microscope the most prominent feature of medullated nerves, affording a nearly complete investment of the nerve axis.

The neurilemma is an elastic connective-tissue envelope, which completely invests the medullary substance. This tubular membrane is nucleated, and at irregular intervals is constricted so as to reach very nearly the axis cylinder. These constrictions are called by Ranvier *nodes*, and it is believed that the perineurium presents a single nucleus between each of these nodal points. The constrictions do not, however, affect the even calibre or continuity of the axis cylinder.

A typical nerve fibril has been described as resembling, structurally, a doubly insulated telegraphic cable, but the comparison is unfortunate and misleading, as the functioning of the nerve bears no resemblance to the phenomena exhibited by electrical conductors.

NERVE CELLS.

Nerve cells are usually grouped, and are the essential feature of nerve centres, otherwise called *ganglia* or gray matter. Ganglion cells are among the largest cell elements of the body, and consist of a dense, reticulated, and frequently pigmented ground work, inclosing a large translucent nucleus, and usually a single nucleolus. One or more prolongations, *poles* or *horns*, are sent from these cells, and hence they have been classified as *unipolar*, *bipolar*, *tripolar*, *quadripolar*, and *multipolar*, according to the number of projections. The cell prolongations generally divide soon after leaving the body, and subdivision continues until exceedingly minute fibrils result, which serve as connecting links of the elements of a ganglion. Usually one (the larger) pole is projected which remains unbranched. This becomes the axis cylinder of a nerve fibril, and affords connection between the elements of a ganglionic centre and the conducting portion of the nervous apparatus.

Ganglion cells are surrounded by irregular channels or lymph-spaces, and are thus in intimate relation with the lymphatic system.

CONNECTIVE TISSUE OF THE NERVOUS SYSTEM.

The connective tissue, which serves to unite the elements of a nerve trunk, does not differ materially from the sustentacular tissue of other organs. Different terms are applied, according to its use and location, as follows:

EPINEURIUM.—*Forming the sheath of the entire nerve trunk.*

PERINEURIUM.—*Surrounding the bundles composing the nerve trunk.*

ENDONEURIUM.—*Permeating and uniting the elements of the bundles.*

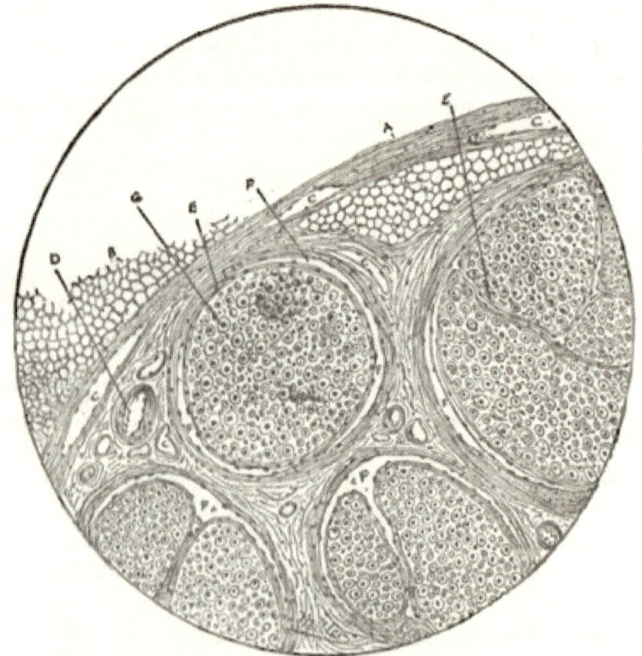

FIG. 119.—TRANSVERSE SECTION OF THE ANTERIOR CRURAL NERVE. × 250.
  A. The epineurium.
  B. Adipose tissue in the loose areolar tissue of the sheath.
  C. Lymph-spaces of the epineurium.
  D. Large blood-vessels of epineurial sheath.
  E. Perineurium surrounding nerve bundles.
  F. Lymph-spaces of last.
  G. Medullated nerves in T. S. supported by connective tissue—endoneurium.

NEURILEMMA.—*Surrounding the individual nerve fibres of a bundle.*

The formula E. P. E. N., composed of the initial of the name of the investments from without inward, will aid the memory.

The epineurium serves to protect the organ in its passage, and to support the nutrient blood-vessels and the channels of lymphatics. The fibres run both longitudinally and transversely. The perineurium, arranged in dense bands, forms distinct sheaths for the nerve bundles, the fibres running, for the most part, circularly. The endoneurium not infrequently divides the nerve bundles into smaller or primitive bundles. It supports the blood capillaries, contains small lymph-spaces, and its nuclei are frequently large and prominent.

The final distribution of the elements of a nerve trunk is effected by subdivision; first, of the large, and afterward of the primitive bundles or fasciculæ. The perineurial sheaths are prolonged, furnishing the dividing bundles, even to the final distribution, where, around terminal and single medullated fibres, the sheath remains as a layer of exceedingly delicate flattened cells. The necessity for the endoneurium ceases with the ultimate subdivision of the nerve fasciculus.

NEUROGLIA.

The sustentacular or supporting tissue of the brain and spinal cord differs materially from ordinary connective tissue. It pre-

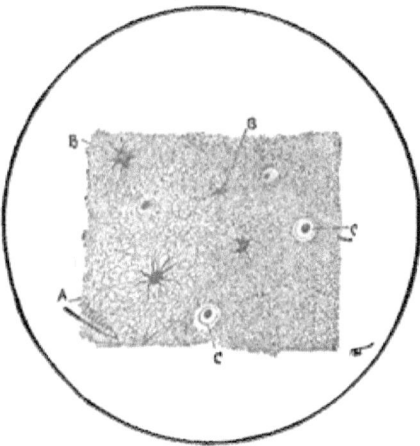

FIG. 120.—NEUROGLIA, FROM BENEATH THE PIA MATER OF THE SPINAL CORD. ×400.

    A. Network of neuroglia fibrils.
    B. Spider (Deiter's) cells.
    C. Nerve fibres in T. S.

sents an interlacement of fibres which, even with the highest powers of the microscope, appear of exceeding tenuity. The neuroglia mesh supports the nervous elements, scattering branched (Deiter's) cells, and small round cells.

Peripheral Termini. (The demonstration of peripheral nerve apparatus should not be attempted until additional work, in the lines hereafter indicated, has secured for the student a degree of perfection in technique which he is not at present supposed to possess.)

SPINAL CORD.

The membranes covering the cord will be discussed later.

The spinal cord is composed of gray (cellular) and white (fibrous) nerve matter, and serves as a medium of communication between the brain and peripheral nerve apparatus. The arrangement of its several parts will be best understood by the study of a transverse section, of which Fig. 121 is a diagrammatic representation.

The gray substance occupies the central portions of the structure, and consists of two lateral masses and a connecting link or commissure. Near the central portion of the figure, a small circu-

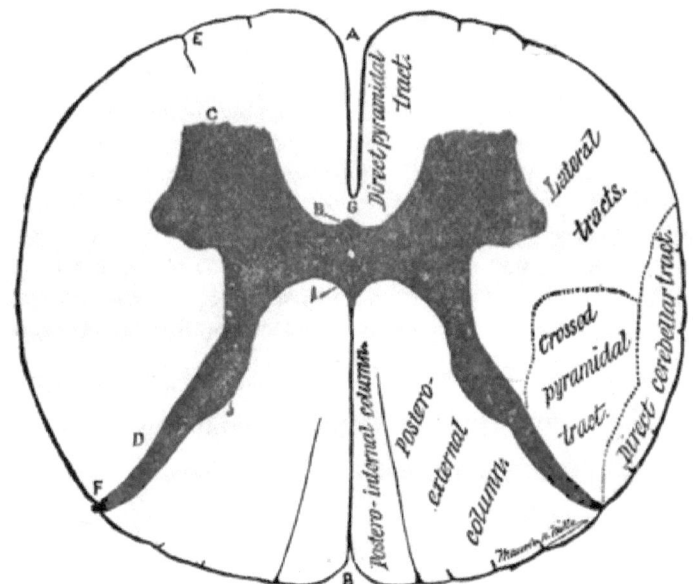

FIG. 121.—DIAGRAM. CERVICAL SPINAL CORD IN TRANSVERSE SECTION.
  A. Anterior median fissure.
  B. Posterior median fissure.
  C. Anterior cornu—gray substance.
  D. Posterior gray cornu.
  E. Point of emergence of anterior root of spinal nerve.
  F. Posterior root of spinal nerve.
  G. White commissure.
  H. Anterior gray commissure.
  I. Posterior gray commissure.
  J. Substantia gelatinosa.

The tracts which are named on the diagram have no definite boundaries histologically. They are physiological areas.

lar opening presents—the transversely divided *central canal*. This is in communication, in the medulla, with the fourth ventricle, and will serve as a starting-point for our study.

The gray matter completely surrounds the central canal, and its outline resembles the capital H. The bars or columns each present anteriorly a blunted extremity, horn or *cornu*, while the posterior cornua are pointed. The lateral bars or columns are connected, as we have seen, a portion of the connecting substance passing in front and a portion behind the central canal—*the anterior and posterior gray commissural bands*.

The white substance is divided anteriorly by the *anterior median fissure*, which sections the cord nearly, but not entirely, to the anterior gray commissure. A corresponding division appears posteriorly (the *posterior median fissure*) which does not divide the cord posteriorly as completely as does the previously-named fissure anteriorly; but the division is indicated by a band of neuroglia, which penetrates entirely to the posterior gray commissure. The two masses of white substance thus indicated by more or less complete central division are termed *lateral white columns*, and these are united just in front of the anterior gray commissure by white nerve tissue—*the white commissure*. The *spinal nerves* take origin from the gray cornua, the *anterior roots* from the anterior and the *posterior roots* from the posterior cornua. The white substance consists essentially of medullated fibres which, with the exception of the anterior spiral nerve roots and the commissural fibres, pass mainly in a longitudinal direction.

PRACTICAL DEMONSTRATION.

Nerve tissue should, under all circumstances, be hardened in Müller's fluid. The cord should be obtained as nearly fresh and uninjured as possible; cut transversely with a sharp razor into pieces half an inch long, and placed immediately in the fluid—in the proportion of a pint of the mixture to two ounces of tissue. The solution should be thrown away after twenty-four hours, and a fresh supply provided. It should be again changed after three days, and again after another week. After four weeks the bichromate should be poured off, and the tissue rinsed once with water; after which the hardening is to be completed with alcohol in the ordinary manner—*i.e.*, commencing with the weak spirit.

After hardening, pieces from the different regions should be cut, and this is best effected by the infiltration methods. Transverse sections are the most instructive, although the student should afterward study longitudinal cuts. The sections must be thin, but not necessarily large, and they may be stained by the method of

Weigert, or with hæma. and eosin. Weigert's method requires very careful manipulation, and is of more special value in pathological research.

If human tissue cannot always be procured in suitable condition, the cord of the ox, pig, sheep, cat, or rabbit will serve well. The ox, especially, provides a means of securing tissue of surpassing excellence, particularly for demonstration of the ganglion cells. The cord of the smaller domestic animals is, in nearly every respect, as valuable for study as that of man, especially as the latter cannot usually be gotten before the serious putrefactive changes, to which nerve tissue is prone, have made marked progress.

HUMAN SPINAL CORD. CERVICAL REGION.

TRANSVERSE SECTION. (Fig. 122.)

OBSERVE:

(L.)

1. General **arrangement of gray and white substance**, with the latter surrounding the former, which resembles in outline the letter H.

2. **Subdivisions of white substance.** (*a*) **Anterior median fissure.** (Note its passage inward and its cessation before reaching the gray substance.) (*b*) **Posterior median fissure.** (Note its shallowness as a true fissure, and the extension of the connective tissue from the bottom inward, until the gray substance is met. Compare the two median fissures.) (*c*) The emergence of the **anterior nerve-roots.** (This provides the external or lateral boundary of anterior white columns or direct pyramidal tracts, the internal boundaries being provided by the anterior median fissure.) (*d*) The **lateral columns.** (These contain the fibres of the crossed pyramidal tract, and include the white substance between the anterior nerve-roots and the posterior gray cornu. Each lateral column contains nerve fibres which pass to the cerebellum—direct cerebellar tract; observe that these tracts have no internal histological boundary. Note the numerous prolongations of the pia mater inward in the lateral columns.) (*e*) The postero-internal or **column of Goll** —*funiculus gracilis*. (These columns present on either side of the posterior median fissure, and are bounded laterally by a prolongation from the pia mater.) (*f*) The **postero-external columns**—*funiculus cuneatus*. (Bounded internally by the postero-internal columns, and externally by the posterior gray cornua.) (*g*) The **white commissure.** (Note the absence of a white commissure posteriorly, the posterior median septum reaching the gray substance.)

3. **Subdivisions of the gray substance.** (*a*) The **central canal.** (Should the section have been taken from the extreme lower cervical cord, this canal as such will be difficult of demonstration, a number of deeply-stained cells only remaining.) (*b*) The **gray commissure,** anterior and posterior. (*c*) The **gray columns.** (*d*) The **anterior gray cornua,** broad and not reaching the periphery of the cord section. (*e*) The **posterior cornua,** narrow and passing completely out, posteriorly, to form the posterior root of a spinal nerve.

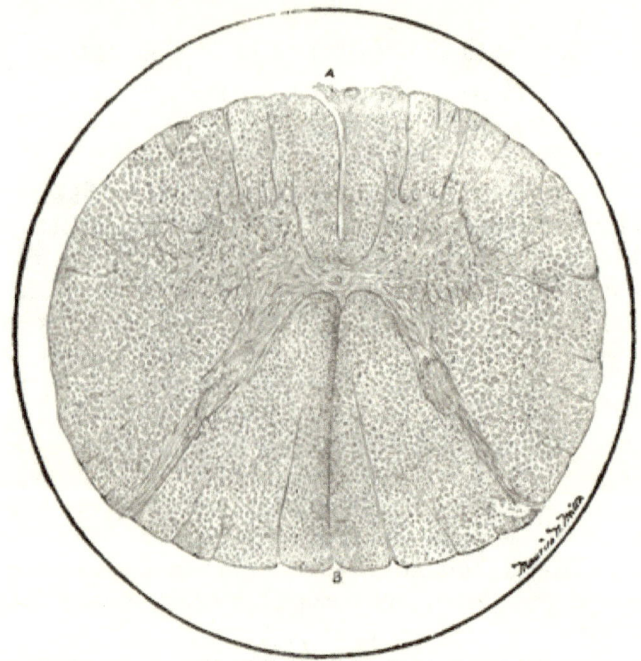

FIG. 122.—TRANSVERSE SECTION OF THE SPINAL CORD. MIDDLE CERVICAL REGION. × 60.
A. Anterior.   B. Posterior.

The references in Fig. 131 apply also to this illustration. Also *vide* text.
This section was made from the cord of a man who died at the age of 75 years, from senile dementia. The gray substance is perfectly normal, but of somewhat diminished area.

(**H.**)

4. The **white substance** (select a field, *e.g.*, in the anterior median column, and observe the transversely-divided nerves). (*a*) The nerves are not collected into fasciculæ, but **each fibre pursues an independent course.** (*b*) The **axis cylinders,** stained lightly with the eosin. (Note the great variation in size.) (*c*) Most of the axis cylinders surrounded by more or less concentric

rings of translucent, unstained **white substance of Schwann.**
(These are medullated fibres.) (*d*) The few and scattering **axis cylinders** without surrounding white substance. (Non-medullated nerves.) (*e*) The **neurilemma**, appearing as a thin, violet ring around the white substance of Schwann. (Most medullated nerves of the cerebro-spinal system are provided with this sheath.) (*f*) The small, about one-three-thousandth of an inch, deeply hæma.-stained **cells of the neuroglia.** (*g*) The **neuroglia substance,** finely granular or fibrillated, between the nerve fibres. (*h*) The **spider cells** (Deiter's) of the neuroglia. (These are not numerous, but easily found near the periphery.) (*i*) The **longitudinal**

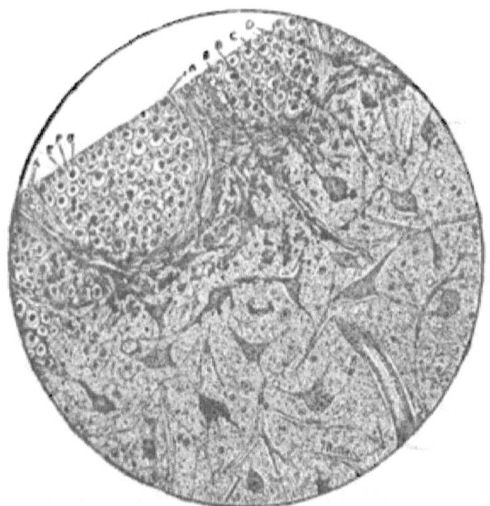

FIG. 123.—SAME SPECIMEN AS SHOWN IN FIG. 122. MORE HIGHLY MAGNIFIED. REGION OF ANTERIOR CORNU. × 350.

A. Medullated filaments passing out from the gray substance to form the anterior root of a spinal nerve.
B. Ganglion cells.
C. Neuroglia nuclei.
D. Blood-vessels.
E. One of the transversely divided medullated fibres of the white substance, anteriorly to the anterior gray cornu. The line leads to the neurilemma.
F. White substance of Schwann—of last.
G. The axis cylinder of E.

nerve fibres passing from the anterior gray cornu to form the anterior root of a spinal nerve. (*j*) The different **size of the nerve fibres in different areas** of the section. (Note the small fibres of the postero-internal column.) (*k*) The **blood-vessels.** (These vessels are largely confined to the neuroglia-septa, which pass in from the pia. These septa contribute to the formation of the nearest approach afforded by the cord of nerve bundles.)

5. The **gray substance**. (*a*) The **central canal**. (The canal is lined with columnar ciliated cells in single layer. The cilia are rarely demonstrable in the human cord, on account of changes which occur very quickly after death. The canal is usually broadest in its lateral diameter, viz., at upper portion of cervical region, from one-one-hundredth to one-two-hundredth of an inch. (*b*) The **ground substance**. (This consists, 1st, of exceedingly minute fibres, formed by the repeated subdivision of the axis cylinders—the **primitive fibrillæ**; 2d, of the delicate **neuroglia fibres**. It is usually difficult in a section to differentiate between the two. Attempts have been made, with more or less success, to differentiate by means of staining agents.) (*c*) **Large ganglion cells**. (Select a field in the anterior horn. The straight, unbranching axis-cylinder process can frequently be distinguished. Note the large, shining nucleus and the deeply-stained nucleolus. These cells are frequently deeply pigmented.) (*d*) **Small ganglion cells**. (Best seen in the posterior horn. In the dorsal cord a collection of medium-sized cells presents, just within the posterior commissure and encroaching upon the white substance posteriorly to this, the column of Lockhart Clarke.) (*e*) **Lymph spaces**. (Observed as a somewhat clear space around the ganglion cells.) (*f*) **Blood-vessels**. (These are much more numerous here than in the white portion; and arteries may frequently be found of considerable size.) (*g*) **Peri-vascular lymphatics**. (Find an artery in transverse section, and observe the clear space around it, which may be mistaken for the result of contraction of the tissue in hardening. Careful study will reveal minute branches of cells, passing between the adventitia of the blood-vessel and the wall of the lymph space.)

THE BRAIN AND ITS MEMBRANES.

The brain and spinal cord are surrounded by three connective-tissue layers—the *dura mater, arachnoid*, and the *pia mater*.

The dura is the most external and the thickest of the three membranes, and constitutes the periosteal lining of the cranial and spinal cavities. It consists largely of elastic tissue, the laminæ and blood-vessels of which are supported by connective tissue. The outer surface is in more or less intimate connection with the bone, and both surfaces are covered with a single layer of thin pavement cells. Beneath is a space—the *subdural*—containing lymph.

The arachnoid, exceedingly thin, presents an outer, glistening, pavement-cell covered surface (isolated from the dura by the subdural space), from the under (inner) side of which short fibrous trabeculæ are projected to the pia. The *subarachnoidal space* is thus seen to consist of numerous communicating chambers, and these spaces are everywhere lined with flat cells, and contain lymph, as does the subdural space.

The pia mater consists of fibrillated connective tissue, usually in intimate connection with the arachnoid externally, by means of the trabeculæ of the latter. The pia is exceedingly vascular, and everywhere covers the brain and cord, and, unlike the arachnoid, penetrates the sulci of the former and the fissures of the latter, becoming continuous with the neuroglia tissue.

The subdural and subarachnoidal spaces are lymph-cavities, and, while not in direct connection one with the other, belong to the general lymphatic system, and are in eventual connection. The two spaces are projected, independently, in the sheaths of the cranial and spinal nerves—the subdural communicating with the lymph channels of the epineurium, and the subarachnoidal with those of the perineurium.

The arrangement of gray and white nerve substance in the brain is precisely the reverse of that of the cord. The gray matter forms an external covering or layer of varying thickness, while the white matter occupies the more central regions. Collections of gray matter—ganglia—are also situate in the deeper parts of the brain substance, the study of which does not come within the limits of this work.

The brain substance does not differ, essentially, from the cord, except in the arrangement of its parts. The nerve fibres are largely medullated, but have no neurilemma. The neuroglia and ganglionic tissues do not here differ in structure from that previously described. The gray substance is arranged in layers, which are, in some instances, quite sharply defined, and again demonstrable only with considerable difficulty.

PRACTICAL DEMONSTRATION.

The tissue is to be prepared in the manner usual with nerve substance—hardened with Müller, followed by alcohol. Thin sections, stained deeply with hæma. and eosin, may be mounted in dammar or, if preferred, in glycerin.

SECTION OF HUMAN CEREBRUM. CUT PERPENDICULARLY TO THE SURFACE. (Fig. 124.)

OBSERVE:

(L.)

1. **The membranes.** (In the drawing only the arachnoid and pia are shown.) (*a*) **The fine fibrillar mesh of the arachnoid.** (*b*) The **nuclei of the flattened cell-covering.** (*c*) The **large blood-vessels.** (*d*) **The pia.** (*e*) Its **continuity with the neuroglia** of the cerebrum.

2. **The outer layer**—the first—of the gray substance. (This layer is poorly defined, but can usually be made out. It consists of primitive nerve fibrillæ, neuroglia fibrils, and scattered spherical cells.)

3. **The second layer.** (This layer presents about the same thickness as the preceding, and will be recognized by the numerous small, triangular nerve-cells. Indeed, these afford the only means of distinguishing the boundary between the two layers, as the stained elements of the outer layer are seldom pyramidal.)

4. **The third layer.** (This layer—the thickest of all the gray laminæ—has been shortened in the drawing, on account of lack of space. It is three or four times as thick as the first layer, and will be readily made out by the large, elongate, pyramidal cells, with their long axes at right angles to the brain surface. Medullated fibres, in more or less distinct bundles, pass between the column-like ganglion cells.)

5. **The fourth layer.** (The large cells of the third layer are seen to stop abruptly, as we pass inward, and give place to small, triangular nerve-cells. This brings us to the fourth plane. Between the cells of this layer, bundles of nerve-fibres are seen, as they radiate toward the cerebral surface.)

6. **The fifth layer.** The line of demarcation between this and the fourth layer is feebly shown; but, on close attention, it will be observed that the small cells of the fourth layer rather abruptly give place to elongate ones, not unlike those of the third stratum. The nerve-bundles are here more plainly indicated.

7. **The white matter.** (The ganglion cells here cease, and the field is occupied with medullated fibres and neuroglia, the

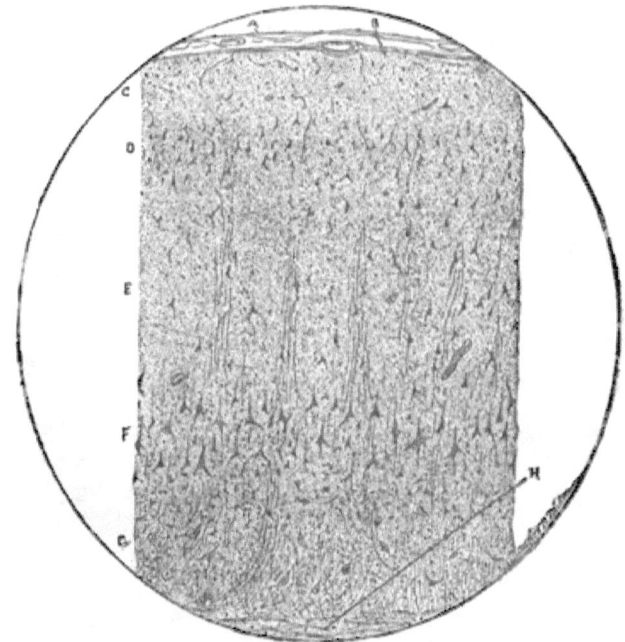

FIG. 124.—VERTICAL SECTION OF CEREBRAL CORTEX. SUPERIOR FRONTAL CONVOLUTION.
× 250.
  A. Arachnoid.
  B. Pia mater.
  C, D, E, F, G. First, second, third, fourth, and fifth layers of gray matter.
  H. White brain substance.

spherical nuclei of the latter becoming prominent from the deep hæma. staining.)

8. **The nutrient blood-vessels.** (The capillaries projected from the pia are especially to be noticed, often of the diameter of a single blood-corpuscle, and presenting as branching lines, composed of these elements—indeed difficult of demonstration when empty. Note the light, perivascular lymph-spaces, well seen around the larger arteries in transverse section.)

VERTICAL SECTION OF HUMAN CEREBELLUM. PREPARED AS LAST SPECIMEN.

(Figs. 125 and 126.)

OBSERVE:

(L.)

1. The **arrangement of the cortex** in the form of leaflets.

2. The **extension of the gray laminæ** within even the minutest folds of the leaves, so as to completely envelop the central

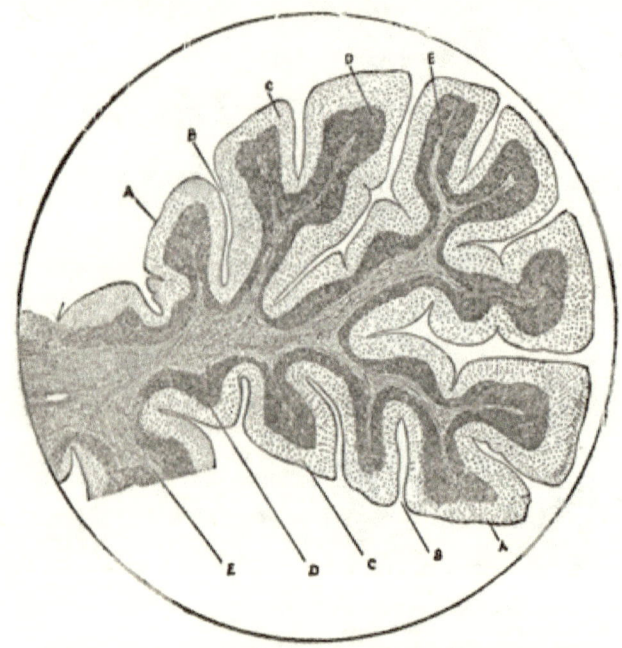

FIG. 125.—LONGITUDINAL SECTION OF ONE OF THE FOLIA OF THE CEREBELLUM. × 60.

    A, A. Line of pia mater.
    B, B. Sulci.
    C, C. Outer layer of gray matter.
    D, D. Inner layer of gray matter, including Purkinje's cells.
    E   White nerve substance.

white nerve-substance. (The staining has been so selected by the tissue as to divide the outer gray matter into two prominent layers. The explanation of this will follow increased amplification.)

3. The **central white matter**. (The fibrillar character can be made out, and the general plan seen to consist, as in the cere-

VERTICAL SECTION OF HUMAN CEREBELLUM. 195

brum, of central nerve fibrillæ radiating toward the cells of the cortical gray substance.)

(H.)

4. The **outer gray layer**. (This is the thickest of the three layers. The prominent elements to be observed are: the **scattering spherical neuroglia and small nerve-cells, nerve-fibrils** passing at right angles to the surface and lost as the outer region is approached, the **prominent blood-vessels** which pass in from the pial investment.)

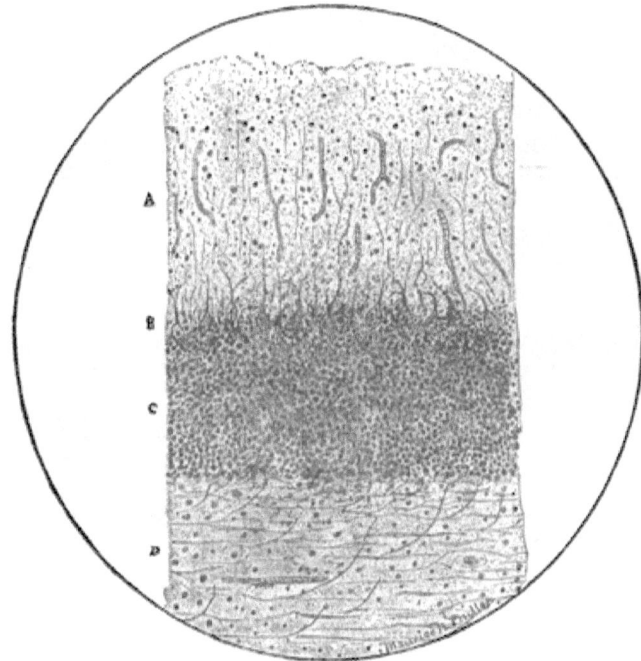

FIG. 126.—VERTICAL SECTION, CORTEX OF CEREBELLUM. PORTION OF SECTION SHOWN IN FIG. 125, MORE HIGHLY MAGNIFIED. × 250.

    A. Outer layer of gray matter.
    B. Layer of Purkinje's cells.
    C. Inner gray layer.
    D. White nerve substance.

5. The **ganglionic layer**. (Directly beneath the outer layer the section becomes deeply stained, from the presence of numerous small cells, among and partly concealed by which are the large **ganglion cells of Purkinje**. These are flask-shaped, and are arranged in a single plane, with their long axes vertically. A

thread-like prolongation may be seen penetrating the layer beneath, providing the cell has been centrally sectioned. **Large horns** are **projected from the outer extremity of the cells,** branches from which provide the nerve-fibrils seen in the last-observed layer. The cells bodies take the eosin, and the nuclei the logwood.)

6. The **granular layer.** (This is the layer seen so distinctly with the low power. It consists of innumerable small, deeply hæma.-stained bodies, usually spherical, which are, as is believed, mostly **neuroglia cells.** These nucleated elements are imbedded in an exceedingly fine matrix of **neuroglia** [Klein] **fibrils.** Search carefully for the axis **cylinder processes of the Purkinje cells** which pierce this layer, and follow them into the white matter below.)

7. **The white substance.** (This consists of medullated nerves which arise largely from the cells of the second gray layer. Klein has also traced fibres into the nuclear layer, and demonstrated their distribution to the small ganglion cells of the lamina, and to the network of the outer gray substance.)

MISCELLANEOUS FORMULÆ.

DAMMAR MOUNTING VARNISH.

No. 1.

Gum dammar,	4 ounces.
Gum mastic (in "tears"),	2 "
Spirits of turpentine,	8 "
Chloroform (Squibb's),	4 "

Mix the dammar with about an equal bulk of clean broken glass. Put in a wide-mouth bottle, and, having added the turpentine, set in a warm place. The glass is added for the purpose of separating the lumps of the resin, and thus hastening the solution. Stir the mixture occasionally.

Add the mastic to the chloroform in a well-stoppered bottle.

When the solution of the resins is complete, add the chloroform mixture to the dammar, and keep in a bottle covered with a plate of glass. After the dirt from the gum has thoroughly settled, decant into small bottles for use. Do not attempt to filter.

I have histological mounts which were made with this medium nearly ten years since, and the tissues have shown no deterioration whatever.

DAMMAR MOUNTING VARNISH.

No. 2.

Best dammar varnish of the paint-shops, diluted with a sufficient quantity of turpentine.

I do not know that this is in any way inferior to the last, but histologists generally have a preference for media of known composition. They may both be diluted with turpentine, chloroform, ether, or pure benzole; or thickened, by removing the stopper of the bottle until a sufficient amount of the solvent has evaporated.

XYLOL BALSAM.

Xylol is preferred by some histologists as a solvent or diluent of Canada balsam for general mounting purposes. Xylol dissolves the balsam very readily without heat, and evaporates more rapidly than most solvents. A thin solution—say xylol two parts to

Canada balsam one part—should be first prepared, and, after filtering through paper, it may be placed in an unstoppered bottle until, by evaporation, it becomes sufficiently thick for use.

Xylol is miscible with strong alcohol, oil of cloves, etc., but not with water.

VARNISHES AND CEMENTS FOR RINGING MOUNTS.

1. *Dammar.*—I believe a clean dammar mount, with circular cover, neatly labelled, cannot be improved in appearance by painting rings of colored varnish around the specimen. Nevertheless, the beginner will purchase and use a turn-table, and must be therefore directed in its employment.

A ring of dammar, thinned with turpentine so as to flow readily from the brush, makes a very neat border to the cover-glass of a specimen mounted in this medium. Let the layer be quite thin.

2. *Zinc Cement.*—To a small quantity of thin dammar varnish, add q. s. of pure dry zinc white. Mix thoroughly on a glass plate with paint-knife or spatula. The consistency should be such as to flow readily from the brush.

Before adding any cement containing pigment or color to a dammar mount, a protecting covering should be applied; otherwise the cement will eventually creep in between the cover and the slide and mix with the original mounting varnish. It may proceed very slowly; but in time the specimen will be surely ruined. This is best avoided by painting a thin ring around the edge of the cover of liquid glue. Cooper's and LePage's are both excellent. Let this dry, and any amount of varnish may be subsequently applied without disaster.

Aniline Colors. Red, blue, etc., may be employed by dissolving the dry color in a little thin shellac varnish. This dries quickly and may be used over the zinc.

Oil colors. The artists' colors which are sold in tubes may be thinned with dammar.

Black varnish. This is the "Black Japan" of the paint shops. It may be made by dissolving gum asphaltum in turpentine. The genuine asphaltum is very difficult to procure, as coal tar is generally sold under this name. It is therefore best to purchase the varnish, thinning with turpentine or pure benzol if necessary.

Shellac Varnish.—The best orange shellac dissolved in strong alcohol, in quantity sufficient to make a varnish of the consistency of treacle.

This is an excellent cement for glycerin mounts. It should be considerably thinned for use as a vehicle for color.

PRESERVATIVE FLUID.

Acetate of soda,	1 pound.
Distilled water,	8 ounces.

Scrapings or teasings from tissues, morphological elements from urinary sediment, casts, in fact, almost any histological element may be indefinitely preserved in this solution. It is important that the tissues, etc., be freed from organic matters in solution, before being placed in the preservative; for example: urine must be carefully decanted from the sediment to be preserved before the acetate solution is added. I am enabled to preserve the different forms of cells from year to year, for my classes, in small bottles of this fluid. When a demonstration is required, it is simply necessary to transfer a drop of the sediment, by means of a pipette, to a slide, and apply the cover-glass.

If it be desired to preserve such a mount, wipe the edges of the cover with a bit of blotting-paper, and seal with a ring of shellac or asphaltum varnish, or zinc cement, applied with a small brush.

NORMAL SALT SOLUTION.

Chloride of sodium (common salt),	7 grains.
Distilled water,	2 fluid ounces.

A medium for the temporary examination of fresh tissues—scrapings, teasings, etc.

RAZOR-STROP PASTE.

Sulphate of iron and common salt equal parts. Calcine in a sand crucible at a dull red heat for ten minutes. When cold, grind lightly in a porcelain mortar and pass through fine gauze. Preserve dry in a well-corked bottle.

A few grains dusted on the surface and mixed with a minute quantity of tallow will add greatly to the efficiency of the strop.

Another.—Equal parts of opticians' rouge and the finest washed flour of emery, mixed with a sufficient quantity of vaseline to make a stiff paste.

SILVER STAINING SOLUTION.

Nitrate of silver,	5 grains.
Water (distilled),	4 ounces.

Used for the demonstration of cement substance between cell elements.

OSMIC ACID SOLUTION.

Osmic acid,	1 gramme.
Water,	100 cubic cent.

The acid is found in market sealed in glass tubes holding one gramme each. The utmost care should be taken to avoid inhaling the vapor from the pure acid as it produces an intense inflammation of the respiratory surfaces. The best method of handling the material is to measure 100 c.c. of water, place it in a strong bottle, and then drop in the tube containing the osmic acid. Then with a glass rod break the tube. The bits of glass need not be removed. Keep the bottle, tightly stoppered, in a cool dark place.

Osmic acid is used in histology on account of its action upon fatty substances—*i.e.*, staining them of a deep brown or black.

WEIGERT'S HÆMATOXYLIN STAINING FOR MEDULLATED NERVE TISSUE.

Formulæ.

1. A saturated aqueous solution of neutral acetate of copper.
2. Strong hæmatoxylin staining fluid, *vide* text.
3. Borax, ferridcyanide of potassium, āā 1 drachm.
  Water, ... 4 fluid ounces.

Process.

1. Allow the sections to remain in 1 for twenty-four hours.*
2. Wash in clean water for a few moments only, and transfer to 2. Brain sections will require from two to three days, and spinal cord twenty-four hours for complete staining. They must appear almost black. If the hardening has been effected with the bichromate solution as given in the text (*vide* hardening fluids), maceration of the sections in the cupric acetate may be omitted.
3. Wash in water for five minutes and transfer to 3. They

---

* The hæma. solution gives the best results, in this staining, if kept at temperature of about 100° F.

may be allowed to remain for twelve hours without harm. Generally an hour will be sufficient.

4. Wash in water.

5. Dehydrate gradually, first placing in dilute alcohol, and afterward in stronger. The sections can now be kept in alcohol indefinitely. If the tissue has been infiltrated with celloidin, the sections must not be held in ninety-five per cent alcohol longer than five minutes, as the infiltrating medium will be dissolved.

6. Clarify with oil of cloves—oil of bergamot for celloidin specimens—and mount in dammar.

The method does not produce bright colors, but it gives a very remarkable differentiation of nerve tissue, by staining the medullary substance violet, and the axis cylinders brown. It is particularly valuable in pathological histology.

BAYBERRY WAX INFILTRATING METHOD.

In answer to many inquiries regarding the material used in this process, I may say that Messrs. Eimer & Amend, chemists, of New York, from whom my supply was originally obtained, state that the article furnished me was the Japan wax. Dr. J. W. Blackburn, of Washington, D. C., kindly informs me that this is the product of *Rhus succedanea*. The material with which I have had the best results was of a very pale yellow or canary color. The darker specimens are unsuitable.

KARYOKINESIS.

The phenomena attending cell-division are best shown in the thin gill-plates or caudal fin of larvæ of the *salamandra*. Very fair demonstrations may be made from rapidly-growing tumors, as *carcinomata*, if after removal they are sliced thin and immediately fixed.

*Flemming's Fixing Fluid.**

Chromic acid,	0.25 per cent	
Osmic acid,	0.1 "	in water.
Glacial acetic acid,	0.1 "	

Half an hour's immersion will usually suffice, after which the tissue is rinsed quickly in water and transferred to absolute alcohol, where it may remain until ready to cut. The parts of larvæ, above mentioned, are of course sufficiently thin without sectioning.

* "Microtomist's Vade-mecum," Lee. London, 1885.

For staining, hæma. or picro-carmine answer well, although saffranin is preferred by Strassburger. A saturated solution of the dye in absolute alcohol is diluted with an equal volume of water and allowed to act on the tissue for twenty-four hours. Wash thoroughly with absolute alcohol, clear in oil of cloves, and mount in dammar.

The highest powers are necessary for successful demonstration.

FIXING AND STAINING THE CORPUSCULAR ELEMENTS OF BLOOD.

The following method, which has been elaborated by Prof. Gaule, of Zurich, will prove more satisfactory than processes which involve the drying of the blood. The essential steps are:

1. *Transferrence to the slide.* This must be accomplished very quickly to provide against changes in the corpuscular elements which occur soon after removal from the vessels. If to be taken from the living animal—and the frog is best for beginners—the surface must be scrupulously clean, a small vessel punctured with the needle, and a minute drop of blood carried to the surface of a slide by means of a glass rod. The blood is spread in a thin layer, after which, and before drying, the elements are to be fixed.

2. *Fixing.*—A portion, say fl. ℥ ij., of a saturated aqueous solution of bichloride of mercury having been prepared in a saucer, the blood slide is submerged in the liquid. Five minutes suffice for this work.

3. *Washing* is accomplished by immersing the slide for a moment in a saucer of distilled water, after which the action is completed by placing in absolute alcohol for five minutes. Drain on bibulous paper for a moment.

4. *Staining.*—Moisten the blood with a little distilled water, drain, and afterward drop a few minims of hæma. solution on the horizontally placed slide. Ordinary hæma. will answer perfectly if, to about a drachm of the solution, two drops of alcohol are added. The staining is complete in five minutes. Wash in distilled water, and again stain as before with a one-per-cent aqueous solution of nigrosin. Again wash with distilled water, and again stain as before with eosin one part, alcohol 50 parts, distilled water 150 parts, for one minute.

5. *Mountin*—A permanent specimen is completed by washing the film of blood with strong alcohol. A drop of oil of cloves gives

translucency in a moment, after which it is drained off, a drop of dammar added, and the cover applied.

The nuclei of the red corpuscle of the frog take the blue hæma., while a variety of white corpuscles with a large round or spindle-shaped nucleus—the *hæmatoblast* of Hayem—has its protoplasm stained blue-black by the nigrosin. Ehrlich has given the name *eosinophilous cells* to those white corpuscles with several nuclei whose granular protoplasm takes the eosin deeply. Other forms of colorless blood-corpuscles as *amœbocytes* and *endothelioid cells* are differentiated by this mode of fixing and triple staining. The highest powers of the microscope are required.

INDEX.

Abbé condenser, 4
Abdominal cavity, 162
    salivary gland, 159
Aberration, chromatic, 2
    spherical, 2
Absorption from intestine, 88
Acetate of copper, 200
Acid, acetic, 31
    chromic, 22
    hydrochloric, 22
    nitric, 22
    osmic, 49
    osmic, solution, 200
    picric, 26
Acini, compound, 155
    simple, 155
Acinous glands, 155
Adenoid reticulum, 167
    tissue, 61, 167
    tissue in thymus body, 178
    tissue, Klein on, 167
Adipose tissue in bronchi, 99
Adventitia of arteries, 66
Afferent glomerular arterioles, 125
Agents, staining, 25
Agminate glands, 91
Ailantus pith, 13
Air, atmospheric, 100
    bubbles, 36
    sacs, 100
    vesicles, 100
Albuminate of silver, 164
Albuminous cement, 163
Alcohol " A," " B," and " C," 21
    hardening, 20
Alum, 25
Alveoli, gland, 155
    pulmonary, 100
Amœbocytes, 203
Ampulliform dilatation, 97
Aniline colors, 198
Anterior commissure, 185
    median fissure, 185
Appendages of skin, 71
Appendix, xiphoid, 58
Arachnoid, 191
Arcade, arterial, 124

Areas, physiological, of spinal cord, 185
Areolar tissue, 51
Arrectores pili, 74
Arrowroot starch, 38
Arterial arcade, 124
Arteries, 66
    adventitia of, 66
    intima of, 66
    large, 67
    lymphatics of, 162
    media of, 66
Arteriolæ rectæ, 126
Artery, bronchial, 95, 98
    hepatic, 107
    phrenic, 150
    pulmonary, 99
    renal, 124, 150
    splenic, 173
    typical, 66
Articular cartilage, 56
Artificial gastric fluid, 22
Ascending limb of Henle, 123
Asphaltum, 198
Atmosphere, 100
Attachment, freezing, 16
Auerbach's plexus, 86, 91
Author's microtome, 15
Axis cylinder, 180

Bacteria in urine, 37
Balsam, xylol, 197
Basement membrane, 99
    membrane of corium, 71
Bayberry tallow, 22, 201
Beaker cells, 97
Bellini, tubule of, 124
Bergamot oil, 27
Bichromate of potash, 22
Bile capillaries, 108
    ducts, origin of, 118
Bipolar cells, 50
    nerve cells, 181
Birds, blood of, 48
Blackburn, Dr. J. W., 201
Black japan, 198
    varnish, 198

Bladder of frog, 62
  urinary, 149
Blood corpuscle as test, 7
  corpuscle, colorless, 49
  corpuscle, red, 47
  corpuscles, staining, 203
  corpuscles, white, 49
  fixing and staining, 202
  of birds, 48
  of camelidæ, 48
  of fishes, 48
  of invertebrates, 48
  of reptiles, 48
  oxygenation of, 100
  plates, 48
  supply, hepatic, 106
  supply of ovary, 149
  supply of spleen, 173
  vessels, 66
  vessels of bronchi, 98
  vessels of epineurium, 182
  vessels of glands, 150
  vessels of kidney, 124
  vessels of omentum, 67
  vessels of pulmonary alveolus, 67
  vessels of skin, 70
  vessels of spinal cord, 189
  vessels, pulmonary, 99
Bodies, Malpighian, of kidney, 122
  Malpighian, of spleen, 173
Body, suprarenal, 120, 150
  thymus, 177
Bone, 58
  corpuscles, 59
  decalcifying, 22
  mounting of, 61
Borax carmine staining fluid, 26
  carmine staining process, 30
Bowman, capsule of, 122
Bowman's muscle discs, 65
Brain, 191
  ganglia of, 191
  gray matter of, 191
  hardening of, 192, 200
  lymph cavities of, 191
  membranes of, 191
  neuroglia of, 182
  section cutting of, 192
  sulci of, 191
  white matter of, 191
Branched tubular glands, 154
Bronchial artery, 98
  glands, 96, 154
  tube, 94
  vein, 98
Bronchi, 94
  adipose tissue in, 99
  basement membrane of, 99
  blood-vessels of, 98
  cartilage in, 98
  glands of, 98

Bronchi, mucosa of, 98
  muscular coat of, 98
  nerves of, 98
  staining of, 97
  subdivision of, 94
  terminal, 97, 101
Bronchus, layers of, 95
  of pig, 44
Brood nests, 179
Brownian movement, 36
Brunner's glands, 89
Bubbles, air, 36
Buccal epithelium, 41

Cable, telegraphic, 181
Calyces of kidney, 120
Camelidæ, 48
Canal, central, 186
Canaliculi, 58
Canal, portal, 108
Canals, dentinal, 78, 81
  Haversian, 59
Cancer, epithelial, 179
Capillaries, 66
  bile, 108
  lymphatic, 161
  of glands, 153
  of ovary, 149
  of supra-renal body, 150
  stomata of, 67
  tortuous, in lung, 103
Capsule of Bowman, 122
  of Glisson, 106
  of kidney, 120
  of lymph nodes, 167
  of thymus body, 177
  supra-renal, 150
Carcinomata, 201
Cardiac gland-tubes, 84
  muscular fibre, 65
Care of objectives, 34
  of the microscope, 34
Carmine and picric acid staining, 31
  movement of particles of, 36
  No. 40, 26
  staining fluid, 26
Carpenter, Prof. Wesley M., 11
Cartilage, 56
  articular, 56
  corpuscles, 56
  elastic, 57
  fibro, 56
  hyaline, 56
  plates in bronchi, 98
  reticular, 57
Casts, urinary, preserving, 199
Caudal fin, 201
Cavities, lymphatic, 161
Cavity, abdominal, 162

INDEX.

Cavity, cranial, 191
    spinal, 191
    thoracic, 162
Cell body, 39
    cement, 163
    distribution, 40
    division, 201
    primordial, 147
    proliferation, 40, 201
    structure, Ehrlich on, 203
    typical, 146
    wall, 39
Celloidin infiltration, 23
Cells, beaker, 97
    bipolar, 50
    bipolar nerve, 181
    border, 85
    central, 85
    chief, 85
    ciliated, 45
    ciliated from oyster, 45
    columnar, 44
    covering, 51
    Deiter's, 183
    endothelial, 51
    endothelioid, 203
    eosinophilous, 203
    epithelial, 51
    flat, 41
    ganglion, 181
    glandular, 51
    goblet, 97
    hepatic, 109
    in urine, 143
    lining, 51
    lymphoid, 61, 161
    mucous, 97
    multinucleated in thymus body, 179
    multipolar, 50
    multipolar nerve, 181
    nerve, 50, 181
    of Purkinje, 194
    outlining of, by silver nitrate, 164
    parietal, 85
    polar, 47, 50
    polyhedral, 49
    quadripolar nerve, 181
    spheroidal, 47
    spider, 183
    stellate, 47, 50
    tailed, from pelvis of kidney, 140
    teased, hepatic, 113
    tripolar, 50
    tripolar nerve, 181
    typical, 39
    unipolar, 50
    unipolar nerve, 181
    uterine, 136
    vacuolated, of bladder, 142

Cells, vaginal, 136
    variation in form, 40
Cement, cell, 163
    zinc, 198
Central canal, 186
Cerebellar tracts, direct, 185
Cerebellum, 194
    cortex of, 194
    folia of, 194
Cerebral layers, 192
Cerebro-spinal system, 180
Cerebrum, practical demonstration of, 192
    staining of, 192
Cervix uteri, glands of, 154
Chains, Graafian, 147
Chamois leather, 34
Change, retrograde, 20
Channels, lymph, 161
    spleen, 173
Chinks, lymphatic, 161
Chloroform, Squibb's, 197
Chromatic aberration, 2
Chromic acid hardening, 21
    acid fixing, 21
Chyle receptacle, 161
Chylopoietic viscera, 106
Cicatricial tissue, 146
Ciliated cells from oyster, 45
    columnar cells, 45
Circulation, lymphatic, 161
Cirrhotic liver, cells from, 52
Cleaning cover-glasses, 31
    slides, 31
Clefts, lymphatic, 161
Clove oil, 28
    oil, removal of, 33
Coarse adjustment, 1
Coiled tubular glands, 154
Collecting tubule, 124
Collodion, 24
Color in air-bubbles, 36
Coloring drawings, 9
Colorless blood-corpuscles, 49
Colors, aniline, 198
    oil, 198
Column of Goll, 187
Columnar cells, 44
Columns, postero-external, 185
    cortical, 120
    postero-internal, 185
Commissure, anterior gray, 185
    posterior gray, 186
    white, 185
Compound acini, 155
    acinous gland, 157
Concavity of blood-discs, 48
Condenser, 4
    Abbé, 4
Conducting portion of nerves, 180
Conductor, electrical, 181
Coniferous wood, 38

Connective-tissue corpuscles, 51
Connective tissue of liver, 106
  tissue of lung, 99
  tissue of nervous system, 182
  tissue of pancreas, 159
  tissue of parotid gland, 157
  tissues, 51
  tissues, special, 61
Conservation of vision, 7
Constrictions of Ranvier, 181
Contractile rods, 63
Convoluted tubule, 122
Cooper's glue, 198
Copper, acetate, 200
  sulphate of, 22
Cords, follicular, 167
Cord, spinal, 185
Corium, 69
  basement membrane of, 71
Cork support, 19
Corn starch, 38
Cornu of spinal cord, 185
Corpuscles, blood, 203
  blood, fixing of, 202
  bone, 59
  cartilage, 56
  connective-tissue, 51
  Hassal's, 179
  lymph, 174
  pus, 49
  salivary, 37, 41
  tactile, 71
Corpus luteum, 144, 146
Cortex of cerebellum, 194
  of kidney, 120
  of thymus body, 178
Cortical columns, 120
Corundum hones, 17
Cotton fibres, 37
Cover-glass, 31
Cover-glass, pressure of, 36
Covering cells, 51
Crossed pyramidal tracts, 185
Crusta petrosa, 80, 81
  petrosa, lacunæ of, 80
Crypts of Lieberkühn, 89
Crystals, fat, 55
Cul-de-sac, vaginal, 136
Currents, thermal, 36
Cuticula, 78
Cylinder, axis, 180

Dammar, mode of using, 33
  varnish, 184
Decalcification, 61
  of teeth, 22, 80
Decalcifying of bone, 22
Degeneration, fatty, 155
Dehydrating tissues, 28
Deiter's cells, 183

Dentinal canals, 78, 82
  elements, 82
  fibres, 78, 82
  sheath, 82
  striæ, 82
Dentine, 78, 81
Deposits, urinary, 135, 143
Derma, 69
  basement membrane of, 71
Descending limb of Henle, 123
Development of ovum, 147
Diameter of kidney tubes, 122, 123
Diaphragm, central tendon of, 163
Differentiation of cell elements, 203
Disc, Hensen's, 63
Discs, intervertebral, 57
  of Bowman, 65
Discus proligerus, 146
Dissociating fluids, 9
  process, 22
Distal convoluted tubule, 123
Distilled water, 163
Distribution, cell, 40
Direct cerebellar tracts, 185
  pyramidal tracts, 185
Division of cells, 201
Dog, kidney of, 127
  stomach of, 86
Double staining, 29, 31
Drainage, tissue, 161
Duct, hepatic, 106
  lymph, 90
  pancreatic, 159
  papillary, 124
Ductless glands, 173
Ducts, lymphatic, 161
  mesenteric, 162
  thoracic, 162
Dura mater, 191
Dust on lenses, 34

Efferent glomerular arteriole, 125
  veins of lymph nodes, 167
Ehrlich on cell elements, 203
Elastic cartilage, 57
  lamina of blood-vessels, 66
  tissue, 52
  tissue in bronchi, 98
Elder pith, 13
Electrical conductor, 181
Elements, dentinal, 82
  of ganglia, 181
  sarcous, 65
  structural, 35
  structural, of nervous system, 180
Embryonic tissue, 62
Emery, flour of, 199
Enamel of teeth, 79
  prisms, 79
Endomysium, 64
Endoneurium, 182

INDEX.

Endothelial cells, 51
Endothelioid cells, 203
Endothelium, 44
    of blood-vessels, 66
Eosin solution, 26
  staining, 203
Eosinophilous cells, 203
E. P. E. N. formula, 182
Epidermis, 68
Epineurium, 182
    lymph-spaces of, 182
    vessels of, 182
Epithelia, glandular, 50
Epithelial cancer, 179
  cells, 51
Epithelium, buccal, 41
    of bladder, 142
    of ovary, 148
    of tongue, 41
    pavement, 42
    proliferation of, 135
    stratified, 41
    squamous, 41
    tessellated, 42
    transitional, 41
    uterine, 136
    vaginal, 136
Ether freezing, 16
Eustachian tube, 58
Extraneous substances, 37
  substances, mounting of, 38
Eye-lens, 3
Eye-piece, 1

Fasciculi, primitive, 64
Fat-columns of Satterthwaite, 70
Fat-crystals, 55
Fat-globules in milk, 35
  in suprarenal capsule, 152
Fat-tissue, 54
Fatty degeneration, 155
Fauces, 158
Feathers, 58
Fenestrated membrane, 66
Fermentation spores, 38
Ferrein, pyramids of, 122, 124
Fibres, cotton, 37
  dentinal, 78, 82
  linen, 37
  nerve, 180
  perforating, 59
  silk, 37
  wool, 37
Fibroblasts, 51
Fibro-cartilage, 56
Fibrous tissue, 51
Field lens, 3
  of view, 5
Fin, caudal, 201
Fine adjustment, 1

Fishes, blood of, 48
Fissure, anterior median, 185
  posterior median, 185
  transverse, of liver, 106
Fixing blood elements, 202
  chromic acid, 21
  fluid, Flemming's, 201
Flat cells, 41
Flour of emery, 199
Fluid, artificial gastric, 22
  dissociating, 9
  Flemming's fixing, 201
  lymphatic, 161
  preservative, 199
Focal adjustment, 5
Focussing, 1, 5
Fœtal life, 147
  lung, 103
  ovary, 148
  teeth, 80
Folia of cerebellum, 194
  of suprarenal bodies, 150
Follicle of hair, 71
Follicles of Lieberkühn, 91
  solitary, 90
  of thymus body, 177
Follicular cords, 167
Foramen dentium, 78
Form of objects, 35
Formulæ, miscellaneous, 197
Fourth ventricle, 186
Freezing attachment, 16
Fresh tissue, 20
Frog, bladder of, 62
Frog's mesentery, 42
Funiculus cuneatus of spinal cord, 187
  gracilis of spinal cord, 187

Gage, Professor, on Japanese paper, 31
Gall-duct, 107
Ganglia, nerve, 181
  of brain, 191
Ganglion cells, 181
  cells, poles of, 181
  elements of a, 181
Gastric fluid, artificial, 22
  juice, 83
  tubules, 83
Gaule, Professor, 202
Gelatinous substance, 185
Genito-urinary tract, 135
Germinal spot, 146
  vesicle, 146
Gill plates, 201
Gland alveoli, 155
  mammary, 155
  parotid, 156
  sebaceous, 73
  serous, 159

Gland sublingual, 157
    submaxillary, 158, 160
    sudoriferous, 71
    thymus, 177
Glands, 153
    acinous, 155
    agminate, 91
    branched tubular, 154
    bronchial, 96, 154
    Brunner's, 89
    buccal, 157
    coiled tubular, 154
    ductless, 173
    essentials of, 153
    gastric, 154
    intestinal, 154
    mucous, 158
    mucous of bronchi, 98
    of cervix uteri, 154
    parenchyma of, 153
    peptic, 83
    Peyer's, 154
    pyloric, 84
    salivary, 153, 157
    sebaceous, 155
    sudoriferous, 154
    uterine, 154
    vessels of, 153
Glandular cells, 50
    epithelia, 50
    structures, staining of, 159
Glassy membrane of hair, 71
Glisson's capsule, 106
Globules, oil, 36
Glomerulus of kidney, 125
Glue, Cooper's, 198
    Le Page's, 198
Glycerin in honing, 17
    in mounting, 128
Goll, column of, 187
Goose-flesh, 74
Graafian chains, 147
    follicles, 144
Granules, pigment, 152
Gray commissure, 185
    matter of brain, 191
    nerve matter, 181
Gum, dammar, 197
    mastic, 197

Hæmatoblast, 202
Hæmatoxylin, 25
    and eosin, 29
    staining fluid, 25
    staining process, 27
    Weigert's, 200
Hair, 37, 71
    cortex of, 91
    follicle, 71
    follicle, muscle of, 74
    glassy membrane of, 71

Hair, from shaving, 74
    medullated, 71
    permanent mounting of, 74
    root-sheath of, 71
    transverse section of, 71
Hardening, alcohol, 20
    bayberry tallow, 22
    by freezing, 24
    chromic acid, 21
    of bladder, 135
    of brain, 192
    of fœtal ovary, 147
    of genito-urinary organs, 135
    of kidney, 127
    of liver, 110, 113
    of tissues, 20
    of nerve fibre, 180
    of os uteri, 135
    of ovary, 144
    of pancreas, 159
    of parotid gland, 159
    of small intestine, 92
    of spinal cord, 186
    of spleen, 174
    of submaxillary gland, 159
    of suprarenal capsule, 150
    of thymus body, 177
    of ureters, 135
    of uterus, 135
    quick, 21
    with Müller's fluid, 22
Hartnack objectives, 4
Hassal's corpuscles, 179
Haversian canals, 59
    system, 59
Hayem's hæmatoblast, 202
Heitzmann, Dr. Carl, 61
    on development of teeth, 80
Henle, loop of, 123
    tubule of, 123
Hensen's middle disc, 63
Hepatic artery, 107
    blood-supply, 106
    cells, 109
    cells, teased, 113
    duct, 106
    lobules, 106
    veins, 106
Hilum of kidney, 120
Hone, 17
    water of Ayr, 17
Horns of ganglion cells, 181
Horny layer of skin, 68
Hyaline cartilage, 55
Hydrocarbons, 88
Hydrochloric acid, 22

Ileum, section of, 92

INDEX.

Illumination, 4
Imbedding with ailantus pith, 13
　　with paraffin, 13
Incremental lines, 79
Infiltrating with bayberry tallow, 22
Infiltration with celloidin, 23
Infundibula of kidney, 120
　　pulmonary, 101
Injection of lung, 101
Insects, 38
Intima of arteries, 66
Interglobular spaces, 78, 82
Interlobular septa of lung, 100
　　veins, 106
　　vessels of kidney, 124
Internal elastic lamina, 66
Intervertebral discs, 57
Intestinal lymphatics, 90
　　villi, 88
Intestine, coats of, 88
　　of rabbit, 45, 92
　　small, 88
Invertebrates, blood of, 48
Involuntary muscle, 63

Japan, black, 198
　　wax, 201
Japanese paper, 31, 34
Juice, gastric, 83

Karyokinesis, 201
Kidney, 120
　　afferent vessels of, 129
　　arterial arcade of, 129
　　blood-vessels of, 124
　　boundary region of, 144
　　Bowman's capsule of, 129
　　calyces of, 120
　　capsule of, 120, 129
　　collecting tubes of, 132
　　convoluted tubes of, 129
　　cortex of, 120
　　cortical labyrinths of, 129
　　diagram of, 121
　　efferent vessels of, 129
　　Ferrein's pyramids of, 129
　　glomerulus of, 130
　　hardening of, 127
　　Henle's limb, 130
　　Henle's loop of, 130
　　hilum of, 120
　　infundibula of, 120
　　interlobular blood-vessels of, 129
　　intertubular capillaries of, 144
　　labyrinths of, 129
　　lymphatics of, 120
　　Malpighian bodies of, 129
　　Malpighian pyramids, 129
　　markings, 129

Kidney, medullary portion of, 133
　　medullary radii of, 129
　　papillary ducts of, 134
　　pelvis of, 120, 140
　　principal tubes of, 134
　　scheme of, 121
　　spiral tubes of, 132
　　tubes, 122
　　tubules, diagram of, 123
　　tubules of, 154
Klein on adenoid tissue, 167
　　on neuroglia, 195
Knives, sharpening, 16
Krause's line, 63

Labels for slides, 34
Laboratory microscope, 2
　　microtome, 15
Labyrinth, kidney, 122
Lacteals, 90
Lacunæ, 58
　　of crusta petrosa, 80
Lamellæ, bone, 59
Lamina, internal elastic, 66
Laminæ of crusta petrosa, 80
Larvæ of salamander, 201
Lateral tracts, 185
Layers of cerebrum, 192
　　of epidermis, 68
Lenses, cleaning soiled, 34
　　water, 36
Le Page's glue, 198
Leucocythæmia, spleen in, 176
Lieberkühn, crypts of, 89
Life, fœtal, 147
Lifting sections, 27
Ligamentum nuchæ, 53
Light, transmitted, 10
Limb of Henle, 123
Limiting membrane, 39
Line, Krause's, 63
　　Purkinje's, 82
Linen fibres, 37
　　for optical surfaces, 34
Lines, incremental, 79
　　of Retzius, 80
Lining cells, 51
　　of stomach, 83
Links, Graafian, 147
Liquor potassæ, 26
Liver, 105
　　connective tissue of, 106
　　lobules of, 106
　　of pig, 116
　　parenchyma, 108, 116
　　physiological scheme, 109
　　practical demonstration, 109
　　scheme of structure, 106
Lobes of thymus body, 177
Lobular parenchyma of liver, 108
Lobule, primary, 100

Lobules, hepatic, 106
    of thymus body, 177
Loop of Henle, 123
Lung, 94
    connective tissue of, 99
    fœtal, 103
    hardening of, 103
    injected, 102
    interlobular septa, 100
    of pig, 103
    parenchyma of, 100
    staining of, 103
    vascular supply of, 100
Lymph, 161
    cavities of brain, 191
    ducts of intestine, 90
    node, diagram of, 168
    node, practical demonstration of, 169
    nodes, 91, 167
    nodes, capsule of, 167
    nodes, sectioning, 169
    nodes, staining, 169
    nodes, trabeculæ of, 167
    nodes, vascular system of, 167
    paths, 164
    paths, valves of, 164
    spaces in portal canals, 116
    spaces of nerves, 182
    spaces of perineurium, 182
    spaces of spinal cord, 190
    spaces of thymus body, 178
Lymphatic capillaries, 161
    ducts, 161
    system, 161
Lymphatics of intestine, 90
    of kidney, 120
    perivascular, 162, 190
Lymphoid cells, 61, 161

Magnification of movements, 36
Magnifying power, 7
Malpighian bodies of spleen, 173
Malpighii, pyramids of, 120
Mammary gland, 155
Markings, kidney, 129
Mastic, 197
Measurement of objects, 7
Media of arteries, 66
Medico-legal work, 103
Medulla, 186
    of hair, 71
    of thymus body, 178
Medullated nerves, 180
Meisser's plexus, 91
Membrana granulosa, 146
    propria of kidney tube, 122
Membrane, basement, 99
    basement of corium, 71
    fenestrated, 66

Membrane, glassy, 71
    limiting, 39
    peridental, 81
    tubular, 181
Membranes of brain, 191
Menopause, 144
Merck's hæmatoxylin, 25
Mesenteric ducts, 25
Mesentery, silvered, 42
Method in observation, 6
    Weigert's, 186
Micrometers, 8
Microscope, 1
    adjustment, 4
    care of, 34
    sketching from, 8
Microtome, laboratory, 15
    Schrauer, 14
    Stirling, 12
    the author's, 15
Milk, fat-globules in, 35
Mirrors, 4, 5
Miscellaneous formulæ, 197
Molecular movement, 37
Mordants, 30
Morphology, Dr. C. Heitzmann's, 80
Mounting objects, 31
    extraneous substances, 38
    varnish, 197
Mounts, rings on, 34
Movement, Brownian, 36
    molecular, 37
    vital, 37
Movements, magnification of, 36
Mucosa of bronchi, 98
    of pelvis of kidney, 140
    of small intestine, 88
    of stomach, 83
    of ureter, 141
    of uterus, 136
    of vagina, 136
Mucous glands, 158
    glands of bronchi, 98
Müller, capsule of, 122
Müller's fluid, 22
    fluid, hardening with, 22
Multinucleated cells of thymus body, 179
Multipolar nerve cells, 181
Muscle, 62
    cardiac, 65
    from tongue, 65
    non-striated, 62
    of hair-follicle, 74
    striated, 63
Muscular coat of bronchi, 98
Muscularis mucosæ of stomach, 83

Nerve cells, 50, 181
    cells, bipolar, etc., 181

Nerve functioning, 181
  ganglia, 181
  spinal, 185
  staining, Weigert's, 200
  trunks, connective tissue of, 181
Nerves, conducting portion of, 180
  of special sense, 181
Network, intracellular, in hepatic cells, 117
Neurilemma, 180
Neuroglia, 61, 181, 186
  Klein on, 194
Nigrosin, 203
Nitrate of silver solution, 26
Nodes, lymph, 167
  of Ranvier, 181
Non-medullated nerves, 181
Non-striated muscle, 62
Normal salt-solution, 180, 199
Nuclei of cells, 39
Nucleoli of cells, 39

Objectives, 1
Objects, form of, 35
  movement of, 36
Odontoblasts, 78, 81
Oil-colors, 198
Oil-globules, 36
Oil of bergamot, 27
  of cloves, 28
Omentum, 55
  vessels from, 67
Optical axis, 5
Optician's rouge, 199
Organisms in urine, 37
Organs, urinary, 135
Origin of bile-ducts, 118
Osmic acid, 49
  acid solution, 200
Osler, Professor, 58
Os uteri, 136
Ova, 146
Ovary, 144
  blood-supply of, 149
  corpus luteum of, 144
  Graafian follicles of, 144
  practical demonstration, 144
Ovum, development of, 147
Ox, spinal cord of, 186
Oxygenation of blood, 100
Oyster, ciliated cells from, 45

Pancreas, 153
  practical demonstration of, 159
Paper, Japanese, Prof. Gage on, 34
Papillary ducts, 124
  eminence, 120
Paraffin cement, 11
  imbedding, 13

Paraffin soldering, 18
Parenchyma, 51
  of glands, 153
  of liver, 108, 116
  of lung, 100
Parotid gland, 156
  gland, parenchyma of, 159
  practical demonstration of, 159
Paste, razor-strop, 199
Patch, Peyer's, 91
Pavement epithelium, 42
Pelvis of kidney, 140
Pepsin, 22
Peptic glands, 83
Perforating fibres of Sharpey, 59
Perichondrium, 96
Peridental membrane, 81
Perimysium, 64
Perineurium, lymph-spaces of, 182
Periosteum, 59
Peripheral nerve termini, 183
Peritoneum, 162
  of stomach and intestines, 83
Perivascular lymphatics, 162, 190
  lymph-spaces of cerebrum, 193
Peyer's patch, 91
Phrenic artery, 150
Pia mater, 191
Picric acid solution, 26
Pig, bronchus of, 45
  kidney of, 127
  liver of, 116
  spinal cord of, 186
Pig's lung, 113
Pigment in suprarenal capsule, 152
Plates, blood, 48
  cartilage, 98
Pleura, 100, 162
Plexus, Auerbach's, 86, 91
  Meissner's, 91
Pencilling of serous surfaces, 162
Polar cells, 47, 50
Poles of ganglion cells, 181
Pollen, 39
Polyhedral cells, 49
Portal canal, 108
  vein, 106
Posterior commissure, 185
Postero-external columns, 185
Postero-internal columns, 185
Potassium ferridcyanide, 200
Potato starch, 38
Power, magnifying, 7
Practical demonstration. Bronchus of pig, 97
  demonstration. Cerebellum, 196
  demonstration. Cerebrum, 192

Practical demonstration. Development of ovary, 147
  demonstration. Human liver, 113
  demonstration. Human lung, 105
  demonstration. Intestine, 92
  demonstration. Kidney, 140
  demonstration. Liver of pig, 109
  demonstration. Lymphatics of central tendon of diaphragm, 163
  demonstration. Mesenteric lymph node, 169
  demonstration. Ovary, 144
  demonstration. Pancreas, 159
  demonstration. Parotid gland, 159
  demonstration. Spinal cord, 186
  demonstration. Spleen, 173
  demonstration. Stomach, 87
  demonstration. Submaxillary gland, 159
  demonstration. Suprarenal capsule, 150
  demonstration. Teeth, 81
  demonstration. Thymus body, 177
  demonstration. Urinary bladder, 141
  demonstration. Ureter, 140
  demonstration. Uterus, 138
  demonstration. Vagina, 138
Preservative fluid, 199
Pressure of cover, 36
Prickle cells of skin, 69
Primitive fasciculi, 64
Primordial cell, 147
Prismatic color in air-bubbles, 36
Prisms, enamel, 79
Process, decalcifying, 22
  dissociating, 22
Processes of odontoblasts, 82
Proliferation, cell, 40, 201
Protoplasm, 89
Proximal convoluted tubule, 122
Pulmonary alveoli, 100
  artery, 99
  infundibula, 101
Pulp of teeth, 78

Pulp of spleen, 173
Purkinje, cells of, 194
  granular layer of, 79
Purkinje's line, 82
Pus-corpuscles, 49
Pyloric glands, 84
Pyramidal tracts, 185
Pyramids of Ferrein, 122
  of Malpighii, 120

Quadripolar nerve cells, 181
Quick hardening, 20

Rabbit, central diaphragmatic tendon of, 163
  intestine, 45, 92
  kidney of, 127
  spinal cord of, 187
Ranvier's nodes, 181
Rapid hardening, 21
Razor stropping, 18
  strop paste, 199
Receptaculum chyli, 91, 161
Red blood-corpuscle, 47
  muscle, 63
Renal artery, 124, 150
Reptiles, blood of, 48
Reservoir, lymph, 161
Rete Malpighii of skin, 69
  mucosum of skin, 69
Reticular cartilage, 57
Reticulum, 40
  adenoid, 167
Retrograde changes, 20
Retzius, lines of, 80
Ringing mounts, 34, 198
Rods, contractile, 63
Rogers' stage micrometer, 8
Root-sheath of hair, 71
Rouge, optician's, 199

Sacs, air, 100
Salamander, 201
Salivary abdominal gland, 154
  corpuscles, 37, 41
  glands, 157
Salt solution, normal, 199
Salter, incremental lines of, 79
Sarcolemma, 63
Sarcous elements, 65
Satterthwaite, fat-columns of, 70
Scalp, hair from, 74
Schrauer microtome, 14
Schwann, white substance of, 180
Sebaceous gland, 74, 155
Sebum, 74
Secretion of succus entericus, 88
  pancreatic, 159
Section cutting, 9
  cutting, free-hand, 10
  cutting, with Stirling microtome, 12

INDEX.

Section lifter, 32
    needle, 27
    spoon, 32
Sediments, urinary, 199
Septa, interlobular, of lung, 100
Septum narium, 56
Serous gland, 159
    surfaces, pencilling of, 163
Sharpening knives, 16
Sharpey's fibres, 59
Sheath, dentinal, 78
Sheep, spinal cord of, 187
Shellac varnish, 198
Silk fibres, 37
Silver, albuminate of, 164
    nitrate of, 26
    solution, 26
    staining solution, 200
Silvered mesentery, 42
Skeletal muscle, 63
Setching from microscope, 8
Skin, 68
    chamois, 34
    practical demonstration, 74
    staining of, 74
    sudoriferous glands of, 72
Slides, labelling, 34
Small intestine, 88
Smooth muscle, 62
Soda, acetate of, 199
Soldering with paraffin, 18
Solitary "glands," 91
    lymphatics, 88
Solution, chromic-acid, 22
    eosin, 26
    normal salt, 180
    osmic-acid, 200
    picric-acid, 26
    silver, 26, 200
Space, subarachnoid, 191
    subdural, 191
Spaces, interglobular, 78, 82
    venous, 173
Special connective tissues, 61
    sense, nerves of, 181
Specimen, permanent, 33
    trays, 34
Spheroidal cells, 47
Spider cells, 183
Spinal cord, 185
    cord, central canal of, 186
    cord, commissures of, 185
    cord, diagram of, 185
    cord, funiculus cuneatus of, 187
    cord, funiculus gracilis of, 187
    cord, lymph-spaces of, 190
    cord, nerve roots of, 185
    cord of domestic animals, 187

Spinal cord, practical demonstration, 186
    cord, physiological areas of, 185
    cord, substantia gelatinosa of, 185
    nerve, 185
Spiral tubule, 122
Spirals from tea leaf, 38
Spleen, 173
    Malpighian bodies of, 173
    practical demonstration of, 173
    pulp, 173
    supernumerary, 174
Spoon, section, 32
Spores, fermentation, 38
    vegetable, 38
Spot, germinal, 146
Squamous epithelium, 41
Squibb's chloroform, 197
Stage micrometer, 8
Staining agents, 25
    carmine and picric acid, 31
    double, 21, 29
    eosin, 203
    fluid, borax-carmine, 26
    fluid, hæma., 25
    fresh tissue, 25
    hæma., 27
    hæma. and eosin, 27
    nigrosin, 203
    osmic acid in nerve tissue, 180, 200
    silver, 163
    solution, silver, 200
    Weigert's nerve, 200
Starch, 38
Stars of Verheyen, 126
Stellate cells, 47, 50
    sternum, 163
Stirling microtome, 12
Stirling's processes, 22
Stomach, 83
    and intestines, 83
    hardening of, 87
    of dog, 86
Stomata, 44, 67, 164
Straight kidney tubule, 124
Stratified epithelium, 41
Stratum corneum of skin, 68
    granulosum of skin, 68
    lucidum of skin, 68
Striæ, dentinal, 82
Striated muscle, 63
Stroma of ovary, 144
Stropping knives, 18
Structural elements, 35
Structure, glandular, 157
Subarachnoid space, 191
Subcutaneous cellular tissue, 70

Subdural space, 191
Sublingual gland, 157
Sublobular veins, 106
Submaxillary gland, 158
Substance, white, of Schwann, 180
Substances, extraneous, 37
Substantia gelatinosa, 185
Succus entericus, 88
Sudoriferous gland, 73, 154
Sulci, 191
Sulphate of copper, 22
Supernumerary spleen, 174
Suprarenal bodies, 120
    capsule, 150
    capsule, practical demonstration of, 150
Sweat-glands, number of, 74
Sweat-tubes, 154
Sympathetic system, 181
System, cerebro-spinal, 180
    Haversian, 59
    lymphatic, 161
    sympathetic nervous, 181
    the nervous, 180
    vascular, of kidney, 125

Tactile corpuscles, 71
Tailed cells from ureter and pelvis of kidney, 140
Tallow, bayberry, 22
Tea leaf, 38
Teased nerve fibres, 180
Teasing, 9
    glandular structures, 159
Teeth, 78
    decalcification of, 80
    development of, 80
    fœtal, 80
    grinding sections of, 80
    practical demonstration of, 80
Telegraphic cable, 181
Tendon, 51
    diaphragmatic, 163
Terminal bronchi, 97, 101
Termini, nerve, 181
Tesselated epithelium, 42
Thermal currents, 36
Thoracic cavity, 162
    duct, 162
Thymus body, hardening of, 177
    body, Hassal's corpuscles in, 179
Tissue, adenoid, 61, 167
    adipose, 54
    adipose in bronchi, 99
    areolar, 51
    cicatricial, 146
    connective, 51
    connective, **of nerve** trunks, 181
    drainage, 161

Tissue, embryonic, 62
    fat, 54
    fibrous, 51
    fresh, 20
    hardening of, 20
    muscular, 62
    sustentacular of brain, 183
    white fibrous, 51
    yellow elastic, 52
Tissues, dehydration of, 28
    selective power of, 25
    special connective, 61
    translucency of, 28
Tongue, 158
    epithelium of, 41
Tooth, canine, 81
Tortuosity of hepatic cell-columns, 115
Trabeculæ of lymph nodes, 167
    splenic, 173
Trachea, 56, 158
Tract, crossed pyramidal, 183
    genito-urinary, 135
Tracts, direct cerebellar, 185
    direct pyramidal, 185
    lateral, 185
Transitional epithelium, 41
Transmitted light, 9
Transverse fissure of liver, 106
Trigone, 142
Tripolar cells, 50
    nerve cells, 181
True skin, 69
Tube, bronchial, 94
    Eustachian, 58
Tubular glands, 153
    membrane, 181
Tubule, Bellini's, 124
    collecting, 124
    gastric, 83
    Henle's, 123
    proximal convoluted, 122
    spiral, 122
    straight, 123
Tubules, uriniferous, 121
Tumors, 201
Tunica albuginea, 144, 147
Turpentine, 197
Typical artery, 66
    cell, 39, 146

Unipolar cells, 50
    nerve cells, 181
Ureter, 120
    mucosa of, 141
Urinary casts, preserving, 199
    deposits, 135
    organs, 135
Urine, bacteria in, 37
    cells in, 143
    course of, in kidney, 122, 126, 127

INDEX.

Uriniferous tubules, 122
Uterus, mucosa of, 136

Vacuolated cells, vaginal, 136
Vagina, cul-de-sac of, 136
    mucosa of, 136
Valves of lymph paths, 164
Variation of cell forms, 40
Varicosities of nerves, 181
Varnish, asphaltum, 198
    black, 198
    dammar, 197
    shellac, 198
Vascular supply of lung, 100
    supply of lymph nodes, 167
    system of spleen, 173
    system of thymus body, 179
Vegetable spores, 38
Vein, bronchial, 98
    portal, 106
Veins, 66
    central, 106
    efferent, of lymph nodes, 167
    hepatic, 106
    interlobular, 106
    intralobular, 106
    sublobular, 106
Venous spaces of spleen, 173
Venulæ rectæ, 126
Venules, 66
Verheyen, stars of, 126

Vesicle, germinal, 146
Vesicles, air, 100
Villi of intestine, 88
Viscera, chylopoietic, 106
Vital movements, 37
Voluntary muscle, 63

Wall, cell, 39
    uterine, 136
Walls of stomach and intestines, 83
Water, boiled, 163
    distilled, 163
    lenses, 36
    of Ayr hone, 17
Wax, bayberry, 22, 201
Weigert's nerve-staining, 186, 200
Wood, coniferous, 38
Wenham objective, 2
Wheat starch, 38
White blood-corpuscles, 49
    commissure, 185
    fibrous tissue, 51
    matter of brain, 191
    substance of Schwann, 180

Xiphoid appendix, 58
Xylol, balsam, 197

Yellow elastic tissue, 52

Zinc cement, 198
Zona pellucida, 146
    vasculosa, 144

www.ingramcontent.com/pod-product-compliance
Lightning Source LLC
Chambersburg PA
CBHW021813230426
43669CB00008B/737